TD 878 .H36 1994
Handbook of bioremediation
189888

DATE DUE

Handbook of
BIOREMEDIATION

NORRIS • HINCHEE • BROWN • MCCARTY
SEMPRINI • WILSON • KAMPBELL
REINHARD • BOUWER • BORDEN • VOGEL
THOMAS • WARD

Project Officer

John E. Matthews
Chief, Applications and Assistance Branch
Robert S. Kerr Environmental Research Laboratory
Ada, Oklahoma

LEWIS PUBLISHERS
Boca Raton   Ann Arbor   London   Tokyo

Library of Congress Cataloging-in-Publication Data

Handbook of bioremediation / by Robert D. Norris ... [et al.] (Robert
    S. Kerr Environmental Research Laboratory).
        p.   cm.
    Includes bibliographical references and index.
    ISBN 1-56670-074-4
    1. Soil remediation.   2. Groundwater--Purification.   3. In situ
    bioremediation.   I. Norris, Robert D.   II. Robert S. Kerr
    Environmental Research Laboratory.
    TD878.H36   1993
    628.5′2--dc20   93-21172
                                                                                              CIP

© 1994 by CRC Press, Inc.
Lewis Publishers is an imprint of CRC Press

No claim to original U.S. Government works
International Standard Book Number 1-56670-074-4
Library of Congress Card Number 93-21172
Printed in the United States of America   1 2 3 4 5 6 7 8 9 0
Printed on acid-free paper

FOREWORD

EPA is charged by Congress to protect the Nation's land, air and water systems. Under a mandate of national environmental laws focused on air and water quality, solid waste management and the control of toxic substances, pesticides, noise and radiation, the Agency strives to formulate and implement actions which lead to a compatible balance between human activities and the ability of natural systems to support and nurture life.

The Robert S. Kerr Environmental Research Laboratory is the Agency's center of expertise for investigation of the soil and subsurface environment. Personnel at the laboratory are responsible for management of research programs to: (a) determine the fate, transport and transformation rates of pollutants in the soil, the unsaturated and the saturated zones of the subsurface environment; (b) define the processes to be used in characterizing the soil and subsurface environment as a receptor of pollutants; (c) develop techniques for predicting the effect of pollutants on ground water, soil, and indigenous organisms; and (d) define and demonstrate the applicability and limitations of using natural processes, indigenous to the soil and subsurface environment, for the protection of this resource.

In-situ bioremediation of subsurface environments involves the use of microorganisms to convert contaminants to less harmful products and sometimes offers significant potential advantages over other remediation technologies. This report provides the most recent scientific understanding of the processes involved with soil and ground-water bioremediation and discusses the applications and limitations of the various in-situ bioremediation technologies.

Clinton W. Hall

Clinton W. Hall
Director
Robert S. Kerr Environmental
Research Laboratory

Dedicated to Richard L. Raymond, Sr.
Through his vision, commitment, and humanity,
he created the
subsurface bioremediation industry.

CONTENTS

Section 3. Bioventing of Petroleum Hydrocarbons

Section 4. Treatment of Petroleum Hydrocarbons in Ground Water by Air Sparging

Section 5. Ground-Water Treatment for Chlorinated Solvents

Section 8. Bioremediation of Chlorinated Solvents Using Alternate Electron Acceptors

Section 9. Natural Bioremediation of Hydrocarbon-Contaminated Ground Water

Section 10.　Natural Bioremediation of Chlorinated Solvents

Section 11.　Introduced Organisms for Subsurface Bioremediation

FIGURES

TABLES

EXECUTIVE SUMMARY

INTRODUCTION

It is the intent of this report to provide the reader with a detailed background of the technologies available for the bioremediation of contaminated soil and ground water. The document has been prepared for scientists, consultants, regulatory personnel, and others who are associated in some way with the restoration of soil and ground water at hazardous waste sites.

The reader is served by this presentation in that it provides the most recent scientific understanding of the processes involved with soil and ground-water remediation, as well as a definition of the state-of-the-art of these technologies with respect to circumstances of their applicability and their limitations. In addition to discussions and examples of developed technologies, the report also provides insights to emerging technologies which are at the research level of formation, ranging from theoretical concepts, through bench scale inquiries, to limited field-scale investigations.

In order for the information in this document to be of maximum benefit, it is important that the reader understand how contaminants are distributed among the various subsurface compartments. This distribution, or phase partitioning of contaminants, is dependent upon a number of factors including the characterization of the contaminants themselves and that of the subsurface environment. This distribution is exemplified in Figure 1 where contaminants are shown to be associated with the vapor phase in the unsaturated zone, a residual phase, or dissolved in ground water.

The report centers around a number of bioremediation technologies applicable to the various subsurface compartments into which contaminants are distributed. The processes which drive these remediation technologies are discussed in depth along with the attributes which direct their applicability and limitations according to the phases into which the contaminants have partitioned. These discussions include in-situ remediation systems, air sparging and bioventing, use of electron acceptors alternate to oxygen, natural bioremediation, and the introduction of organisms into the subsurface. The contaminants of major focus in this report are petroleum hydrocarbons and chlorinated solvents.

IN-SITU BIOREMEDIATION OF SOIL AND GROUND WATER

Bioremediation of excavated soil, unsaturated soil, or ground water (Figure 2) involves the use of microorganisms to convert contaminants to less harmful species in order to remediate contaminated sites. In order for these biodegradative processes to occur, microorganisms require the presence of certain minerals, referred to as nutrients, and an electron acceptor. Several other conditions, i.e. temperature, pH, etc., impact the effectiveness of these processes. The use of biooxidation for environmental purposes has existed for many years and has led to considerable information regarding the biodegradability of specific classes of compounds, nutrient and electron acceptor requirements, and degradation mechanisms. Activated sludge and other suspended growth systems have been used for decades to treat industrial and municipal wastes. Land treatment processes for municipal wastewater and petroleum refinery and municipal wastewater sludges have also been practiced for several decades and have generated a great deal of information

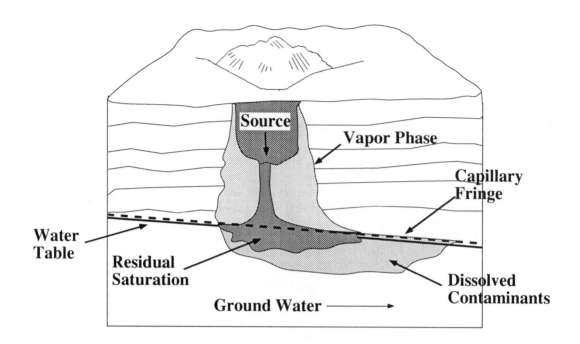

Figure 1.   Distribution of contaminants in the subsurface.

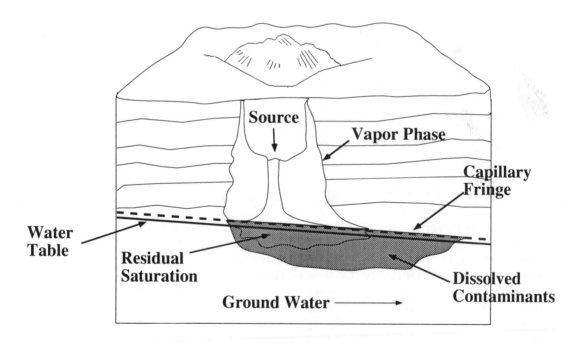

Figure 2.   Contaminant locations treated by in-situ bioremediation.

on nutrient requirements, degradation rates, and other critical parameters affecting biological oxidation.

In the 1970s, tests were conducted to evaluate biological degradation of petroleum hydrocarbons in aquifers. Results from these tests demonstrated that in-situ bioremediation could reduce levels of hydrocarbons in aquifers, and provided considerable information concerning the processes which take place and the requirements necessary to drive these processes.

Although a variety of minerals are required by the microorganisms, it is usually necessary to add only nitrogen and phosphorus. The most common electron acceptor used in bioremediation is oxygen. Stoichiometrically, approximately three pounds of oxygen are required to convert one pound of hydrocarbon to carbon dioxide. Nutrient requirements are less easily predicted. If all hydrocarbons are converted to cell material, however, it can be assumed that nutrient requirements of carbon to nitrogen to phosphorus ratios are in the order of 100:10:1. In some cases where the levels of contaminants are low, sufficient nitrogen and phosphorus are naturally present, and only oxygen is required for the biological processes to proceed.

In-situ bioremediation systems for aquifers typically consist of extraction points such as wells or trenches, and injection wells or infiltration galleries. In most cases, the extracted ground water is treated prior to the addition of oxygen and nutrients, followed by subsequent reinjection.

Critical to the design of an in-situ bioremediation system is the ground-water flow rate and flow path. The ground-water flow must be sufficient to deliver the required nutrients and oxygen according to the demand of the organisms, and the amended ground water should sweep the entire area requiring treatment. This is a critical point in that it is often the hydraulic conductivity of the ground-water system itself or the variability of the aquifer materials which limits the effectiveness of in-situ technologies or prevents its utility entirely. A suggested target for in-situ remediation technologies is a hydraulic conductivity of at least 10^{-4} cm/sec (100 ft/yr). The results of a number of referenced studies suggest that in-situ bioremediation of the subsurface is usually limited to formations with hydraulic conductivities of 10^{-4} cm/sec (100 ft/yr) or greater to overcome the difficulty of pumping fluids through contaminated formations.

In-situ bioremediation systems are often integrated with other remediation technologies either sequentially or simultaneously. For example, if free phase hydrocarbons are present, a recovery system should be used to reduce the mass of free phase product prior to the implementation of bioremediation. In-situ vapor stripping can be used to both physically remove volatile hydrocarbons and to provide oxygen for bioremediation. These systems can also reduce levels of residual phase hydrocarbons as well as constituents adsorbed to both unsaturated soils and soils which become unsaturated during periods when the water table is lowered.

As a class, petroleum hydrocarbons are biodegradable. The lighter soluble members are generally biodegraded more rapidly and to lower residual levels than are the heavier, less soluble members. Thus monoaromatic compounds such as benzene, toluene, ethylbenzene, and the xylenes are more rapidly degraded than the two-ring compounds such as naphthalene, which are in turn more easily degraded than the three-, four-, and five-ring compounds.

Polyaromatic hydrocarbons are present in heavier petroleum hydrocarbon blends and particularly in coal tars, wood treating chemicals, and refinery waste sludges. These compounds have only limited solubility in water, adsorb strongly to soils, and degrade at rates much slower than monoaromatic hydrocarbons.

Nonchlorinated solvents used in a variety of industries are generally biodegradable. For example, alcohols, ketones, ethers, carboxylic acids, and esters are readily biodegradable but may be toxic to the indigenous microflora at high concentrations due to their high water solubility.

Lightly chlorinated compounds such as chlorobenzene, dichlorobenzene, chlorinated phenols, and the lightly chlorinated PCBs are typically biodegradable under aerobic conditions. The more highly chlorinated analogs are more recalcitrant to aerobic degradation but more susceptible to degradation under anaerobic conditions.

Chlorinated solvents and their natural transformation products represent the most prevalent organic ground-water contaminants in the country. These solvents, consisting primarily of chlorinated aliphatic hydrocarbons, have been widely used for degreasing aircraft engines, automotive parts, electronic components, and clothing.

In-situ biodegradation of most of these solvents depends upon cometabolism and can be carried out under aerobic or anaerobic conditions. Cometabolism requires the addition of an appropriate primary substrate to the aquifer and perhaps an electron acceptor, such as oxygen or nitrate, for its oxidation.

In the early 1980s there were few companies that had experience in the bioremediation of soil and ground water. Since that time many companies have utilized bioremediation technologies, although claims of experience are frequently overstated. There now exists a number of organizations and specialists that are knowledgeable in the field of in-situ bioremediation. Several environmental companies have staffs that are experienced in the application of this technology. Many large corporations, especially the oil and chemical companies, have also developed in-house expertise. Some of the U.S. Environmental Protection Agency laboratories as well as Department of Defense and Department of Energy groups have conducted laboratory research and field demonstration studies concerning bioremediation.

BIOVENTING

Bioventing is the process of supplying air or oxygen to soil to stimulate the aerobic biodegradation of contaminants. This technology is applicable to contaminants in the vadose zone and contaminated regions of an aquifer just below the water table (Figure 3). This in-situ process may be applied to the vadose zone as well as an extended unsaturated zone caused by dewatering. Bioventing is a modification of the technology referred to as soil vacuum extraction, vacuum extraction, soil gas extraction, and in-situ volatilization.

Laboratory research and field demonstrations involving bioventing began in the early 1980s, with particular emphasis to the remediation of soil contaminated with hydrocarbons. Early on, researchers concluded that venting would not only remove gasoline by physical means, but would also enhance microbial activity and promote the biodegradation of gasoline. Much of the success of this technology is because the use of air as a carrier of oxygen is 1,000 times more efficient than water. It is estimated that various forms of bioventing have been applied to more than 1,000 sites worldwide, however, little effort has been given to the optimization of these systems.

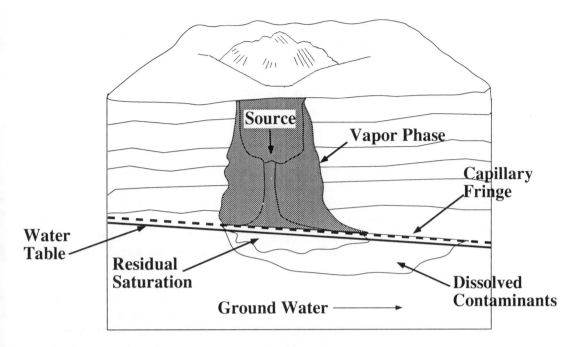

Figure 3. Contaminant locations treated by bioventing.

Bioventing is potentially applicable to any contaminant that is more readily biodegradable aerobically than anaerobically. Although most applications have been to petroleum hydrocarbons, applications to PAH, acetone, toluene, and naphthalene mixtures have been reported. In most applications, the key is biodegradability versus volatility. If the rate of volatilization significantly exceeds the rate of biodegradation, removal essentially becomes a volatilization process.

In general, low-vapor pressure compounds (less than 1 mm Hg) cannot be successfully removed by volatilization, and can only be biodegraded in a bioventing application. Higher vapor pressure compounds (above 760 mm Hg) are gases at ambient temperatures and therefore volatilize too rapidly to be biodegraded in a bioventing system. Within this intermediate range (1 - 760 mm Hg) lie many of the petroleum hydrocarbon compounds of regulatory interest, such as benzene, toluene, and the xylenes, that can be treated by bioventing.

In addition to the normal site characterization required for the implementation of this or any other remediation technology, additional investigations are necessary. Soil gas surveys are required to determine the amount of contaminants, oxygen, and carbon dioxide in the vapor phase; the latter are needed to evaluate in-situ respiration under site conditions. An estimate of the soil gas permeability along with the radius of influence of venting wells is also necessary to design full-scale systems, including well spacing requirements, and to size blower equipment.

Although bioventing has been performed and monitored at several field sites, many of the effects of environmental variables on bioventing treatment rates are still not well understood. In-situ respirometry at additional sites with drastically different geologic conditions has further defined environmental limitations and site-specific factors that are pertinent to successful bioventing. However, the relationship between respirometric data and actual bioventing treatment rates has not been clearly determined. Concomitant field respirometry and closely monitored field bioventing studies are needed to determine the type of contaminants that can successfully be treated by in-situ bioventing and to better define the environmental limitations to this technology.

AIR SPARGING

Air sparging is the injection of air under pressure below the water table to create a transient air-filled porosity by displacing water in the soil matrix. Air sparging is a remediation technology applicable to contaminated aquifer solids and vadose zone materials (Figure 4). This is a relatively new treatment technology which enhances biodegradation by increasing oxygen transfer to the ground water while promoting the physical removal of organics by direct volatilization. Air sparging has been used extensively in Germany since 1985 but was not introduced to the United States until recently.

When air sparging is applied, the result is a complex partitioning of contaminants between the adsorbed, dissolved, and vapor states. Also, a complex series of removal mechanisms are introduced, including the removal of volatiles from the unsaturated zone, biodegradation, and the partitioning and removal of volatiles from the fluid phase. The mechanisms responsible for removal are dependent upon the volatility of the contaminants. With a highly volatile contaminant, for example, the primary partitioning is into the vapor phase, and the primary removal mechanism is through volatilization. By contrast, contaminants of low volatility partition into the adsorbed or dissolved phase, and the primary removal mechanism is through biodegradation.

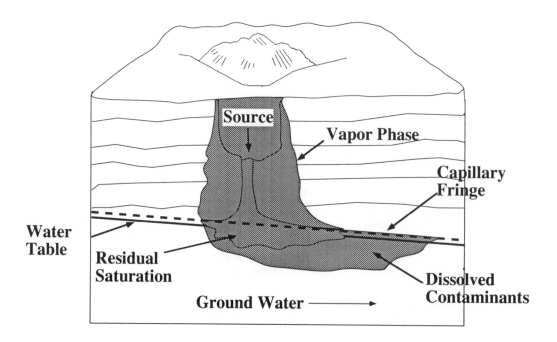

Figure 4.   Contaminant locations treated by air sparging.

One of the problems in applying air sparging is controlling the process. In either bioventing or ground-water extraction, the systems are under control because contaminants are drawn to the point of collection. By contrast, air sparging systems cause water and contaminants to move away from the point of injection which can accelerate and aggravate the spread of contamination. Changes in lithology can profoundly affect both the direction and velocity of air flow. A second problem in air sparging is accelerated vapor travel. Since air sparging increases the vapor pressure in the vadose zone, any exhausted vapors could be drawn into receptors such as basements. As a result, in areas with potential vapor receptors, air sparging should be done with vent systems which allow an effective means of capturing sparged gases.

As with any technology, there are limitations to the utility and applicability of air sparging. The first is associated with the type of contaminants to be removed. For air sparging to work effectively, the contaminant must be relatively volatile and relatively insoluble. If the contaminant is soluble and nonvolatile, it must be biodegradable. The second limitation to the use of air sparging is the geological character of the site. The most important geological characteristic is the homogeneity of the site. If significant stratification is present, there is a danger that sparged air could be held below an impervious layer and spread laterally, thereby resulting in the spread of contamination.

Another constraint of concern is depth related. There is both a minimum and maximum depth for a sparge system. A minimum depth of 4 feet, for example, may be required for a sufficient thickness to confine the air and force it to "cone-out" from the injection point. A maximum depth of 30 feet might be required from the standpoint of control. Depths greater than 30 feet make it difficult to predict where the sparged air will travel.

ALTERNATE ELECTRON ACCEPTORS

Bioremediation using electron acceptors other than oxygen is potentially advantageous for overcoming the difficulty in supplying oxygen for aerobic processes. Nitrate, sulfate, and carbon dioxide are attractive alternatives to oxygen because they are more soluble in water, inexpensive, and nontoxic to microorganisms. The demonstration of this technology in the field is limited, therefore, its use as an alternate electron acceptor for bioremediation must be viewed as a developing treatment technology. Figure 5 illustrates the location of contaminants that may be remediated by introduction of alternate electron acceptors.

Some compounds are only transformed under aerobic conditions, while others require strongly reducing conditions, and still others are transformed in both aerobic and anaerobic environments. In the absence of molecular oxygen, microbial reduction reactions involving organic contaminants increase in significance as environmental conditions become more reducing. In this environment, some contaminants are reduced by a biological process known as reductive dehalogenation. In reductive dehalogenation reactions, the halogenated compound becomes the electron acceptor. In this process, a halogen is removed and is replaced with a hydrogen atom.

Bioremediation with alternate electron acceptors involves the stimulation of microbial growth by the perfusion of electron donors, electron acceptors, and nutrients through the formation. Addition of alternate electron acceptors other than nitrate for bioremediation has not been documented at field scale but has been widely studied at laboratory scale. Nitrate as an electron

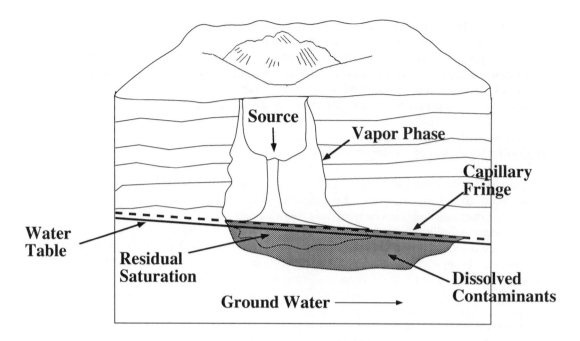

Figure 5.   Contaminant locations treated by alternate electron acceptors.

acceptor has been used for bioremediation of benzene, toluene, ethylbenzene, and xylenes in ground water and on aquifer solids. As for other in-situ remediation technologies, formations with hydraulic conductivities of 10^{-4} cm/sec (100 ft/sec) or greater are most amenable to bioremediation.

The combination of an anaerobic process followed by an aerobic process has promise for the bioremediation of highly chlorinated organic contaminants. Generally, anaerobic microorganisms reduce the number of chlorines on a chlorinated compound via reductive dechlorination, and susceptibility to reduction increases with the number of chlorine substitutes. Conversely, aerobic microorganisms are more capable of transforming compounds with fewer chlorinated substitutes. With the removal of chlorines, oxidation becomes more favorable than does reductive dechlorination. Therefore, the combination of anaerobic and aerobic processes has a potential utility as a control technology for chlorinated solvent contamination.

NATURAL BIOREMEDIATION

The basic concept behind natural bioremediation is to allow naturally occurring microorganisms to degrade contaminants that have been released into the subsurface. It is not a "no action" alternative, as in most cases it is used to supplement other remediation techniques. In some cases, only the removal of the primary source may be necessary. In others, conventional ground-water remediation techniques such as pump and treat may be used to reduce contaminant concentrations within the aquifer.

Natural bioremediation is capable of treating contaminants aerobically in the vadose zone and at the margins of plumes (Figure 6), where oxygen is not limiting. Some sites have shown that anaerobic bioremediation processes also occur naturally and can significantly reduce contaminant concentration on aquifer solids and in ground water. Benzene, toluene, ethylbenzene, and xylene can be removed anaerobically in methanogenic or sulfate-reducing environments; highly chlorinated solvents can undergo reductive dechlorination in anaerobic environments.

While there are no "typical" sites, it may be helpful to consider a hypothetical site where a small release of gasoline has occurred from an underground storage tank (Figure 7). Rainfall infiltrating through the hydrocarbon-contaminated soil will leach some of the more soluble components including benzene, toluene, and xylenes. As the contaminated water migrates downward through the unsaturated zone, a portion of the dissolved hydrocarbons may biodegrade. The extent of the biodegradation will be controlled by the size of the spill, the rate of downward movement, and the appropriateness of requisite environmental conditions. Dissolved hydrocarbons that are not completely degraded in the unsaturated zone will enter the saturated zone and be transported downgradient within the water table where they will be degraded by native microorganisms to an extent limited by available oxygen and other subsurface conditions. The contaminants that are not degraded will move downgradient under anaerobic conditions. As the plume migrates, dispersion will mix the anaerobic water with oxygenated water at the plume fringes. This is the region where most natural aerobic degradation occurs.

One of the major factors controlling the use of natural bioremediation is the acceptance of this approach by regulators, environmental groups, and the public. The implementation of these systems differs from conventional techniques in that a portion of the aquifer is allowed to remain

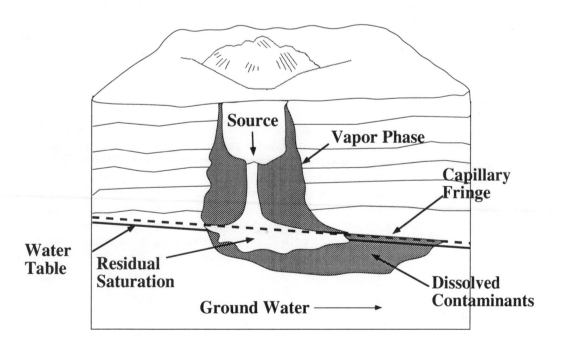

Figure 6.   Contaminant locations treated by aerobic natural bioremediation.

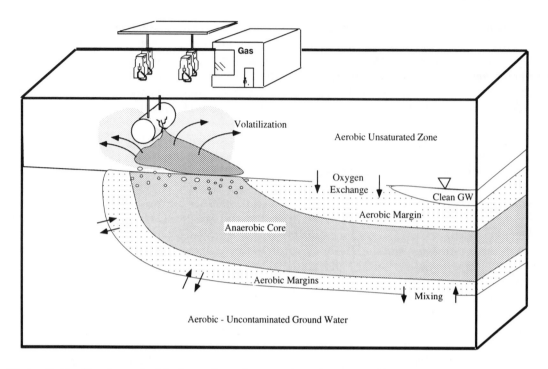

Figure 7. Profile of a typical hydrocarbon plume undergoing natural bioremediation.

contaminated. This results in the necessity of obtaining variances from regulations, and some type of risk evaluation is usually required. Even when public health is not at risk, adjoining land owners may have strong concerns about a contaminant plume migrating under and potentially impacting their property. Therefore, control of plume migration at these sites, usually utilizing some type of hydraulic system, is often necessary. Although natural bioremediation imposes few costs other than monitoring and the time for natural processes to proceed, the public may perceive that this is a "no action" alternative. These various factors may generate opposition to selecting natural bioremediation rather than conventional technologies.

There is almost no operating history to judge the effectiveness of natural bioremediation. In addition, there are currently no reliable methods for predicting its effectiveness without first conducting extensive field testing. This is often the primary reason why natural bioremediation is not seriously considered when evaluating remedial alternatives. At many low priority sites, regulators may have assumed that natural bioremediation would control the migration of dissolved contaminants. Often, these sites have not been adequately characterized nor have they been monitored to determine the effectiveness of this remediation technology. At present, there are no well-documented, full-scale investigations of natural bioremediation, but there is a considerable amount of ongoing research concerning the processes which drive this potentially effective remediation alternative.

INTRODUCED ORGANISMS

Historically, the movement of microorganisms in the subsurface was first discussed in the mid-1920s in relation to the enhanced recovery of oil by the production of biological surfactants and gases. Later, the transport of bacteria through soil was studied to measure the effectiveness of soil-based sewage treatment facilities such as pit latrines and septic tanks in terms of the removal of pathogens. In recent years, research has been directed toward the introduction of microorganisms to soil and ground water to introduce specialized metabolic capabilities, to degrade contaminants which resist the degradative processes of indigenous microflora, or when the subsurface has been sterilized by contaminants. In these attempts to introduce microorganisms to the subsurface, it is often difficult to differentiate their activities from indigenous populations. The use of introduced microorganisms has proven most successful in surface bioreactors when treating extracted ground water in closed-loop recirculation systems.

For added organisms to be effective in contaminant degradation, they must be transported to the zone of contamination, attach to the subsurface matrix, survive, grow, and retain their degradative capabilities. There are a number of phenomena which affect the transport of microbes in the subsurface including grain size, cracks and fissures, removal by sorption in sediments high in clay and organic matter, and the hydraulic conductivity. Many other factors affect the movement of microorganisms in the subsurface including their size and shape, concentration, flow rate, and survivability.

The use of microorganisms with specialized capabilities to enhance bioremediation in the subsurface is an undemonstrated technique. However, research has been conducted to determine the potential for microbial transport through subsurface materials, public health effects, and microbial enhanced oil recovery.

SUMMARY

This report has been prepared by leading soil and ground-water remediation scientists in order to present the latest technical, institutional, and cost considerations applicable to subsurface remediation systems. It is aimed at scientists, consultants, regulatory officials, and others who are, in various ways, working to achieve efficient and cost-effective remediation of contaminants in the subsurface environment.

The document contains detailed information about the processes, applications, and limitations of using remediation technologies to restore contaminated soil and ground water. Field tested as well as new and innovative technologies are discussed. In addition, site characterizations requirements for each remediation technology are discussed along with the costs associated with their implementation. A number of case histories are presented, and knowledge gaps are pointed out in order to suggest areas for which additional research investigations are needed.

SECTION 1

1.1. INTRODUCTION

The purpose of this report is to provide the reader with a detailed background of the fundamentals involved in the bioremediation of contaminated surface soils, subsurface materials, and ground water. A number of bioremediation technologies are discussed along with the biological processes driving those technologies. The application and performance of these technologies are also presented. These discussions include in-situ remediation systems, air sparging and bioventing, use of electron acceptors alternate to oxygen, natural bioremediation, and the introduction of organisms into the subsurface. The contaminants of major focus in this report are petroleum hydrocarbons and chlorinated solvents.

Location of the contamination in the subsurface is critical to the implementation and success of in-situ bioremediation. Also important to success is the chemical nature and physical properties of the contaminant(s) and their interactions with geological materials.

In the unsaturated zone, contamination may exist in four phases (Huling and Weaver, 1991): (1) air phase - vapor in the pore spaces; (2) adsorbed phase - sorbed to subsurface solids; (3) aqueous phase - dissolved in water; and (4) liquid phase - nonaqueous phase liquids (NAPLs). Contamination in saturated material can exist as residual saturation sorbed to the aquifer solids, dissolved in the water, or as a NAPL. Contaminant transport occurs in the vapor, aqueous, and NAPL phases. The interactions between the physical properties of the contaminant influencing transport (density, vapor pressure, viscosity, and hydrophobicity) and those of the subsurface environment (geology, aquifer mineralogy and organic matter, and hydrology) determine the nature and extent of transport.

NAPL existing as a continuous immiscible phase has the potential to be mobile, resulting in widespread contamination. Residual saturation is the portion of the bulk liquid retained by capillary forces in the pores of the subsurface material; the NAPL is no longer a continuous phase but exists as isolated, residual globules. Residual phase saturation will act as a continuous source of contamination in either saturated or unsaturated materials due to dissolution into infiltrating water or ground water, or volatilization into pore spaces.

Liquids less dense than water, such as petroleum hydrocarbons, are termed light nonaqueous phase liquids (LNAPLs). LNAPLs will migrate vertically until residual saturation depletes the liquid or until the capillary fringe is reached (Figure 1.1). Some spreading of the bulk liquid will occur until the head from the infiltrating liquid is sufficient to penetrate to the water table. The hydrocarbons will spread laterally and float on the surface of the water table, forming a mound that becomes compressed into a spreading lens due to upward pressure of the water (Hinchee and Reisinger, 1987). Fluctuations of the water table due to seasonal variations, pumping, or recharge can result in movement of bulk liquid further into the subsurface with significant residual contamination present beneath the water table. The more soluble constituents will dissolve from the bulk liquid into the water and will be transported with the migrating ground water.

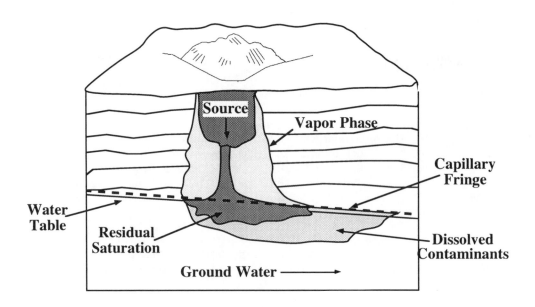

Figure 1.1.   Distribution of petroleum hydrocarbons in the subsurface.

Vertical migration of dense nonaqueous phase liquids (DNAPLs) will continue through soils and unsaturated materials under the forces of gravity and capillary attraction until the capillary fringe or a zone of lower permeability is reached. The bulk liquid spreads until sufficient head is reached for penetration into the capillary fringe to the water table. Because the density of chlorinated solvents is greater than that of water, DNAPLs will continue to sink within the aquifer until an impermeable layer is reached (Figure 1.2). The chlorinated solvents will then collect in pools or pond in depressions on top of the impermeable layer. DNAPL contamination in heterogeneous subsurface environments (Figure 1.3) is difficult to both identify and remediate.

Bioremediation of ground waters, aquifer solids, and unsaturated subsurface materials is widely practiced for contaminants derived from petroleum products. Currently, the most important techniques for bioremediating petroleum-derived contaminants are based on enhancement of indigenous microorganisms by delivery of an appropriate electron acceptor plus nutrients to the subsurface. These techniques are in-situ bioremediation, bioventing, and air sparging; natural bioremediation of petroleum hydrocarbons is also discussed. This paper presents sections devoted to each of the above-mentioned techniques authored by experts actively engaged in bioremediation and research. The sections are: Section 2, *In-situ Bioremediation of Soils and Ground Water Contaminated With Petroleum Hydrocarbons*; Section 3, *Bioventing of Petroleum Hydrocarbons*; Section 4, *Treatment of Petroleum Hydrocarbons in Ground Water By Air Sparging*; Section 7, *In-situ Bioremediation Technologies for Petroleum-Derived Hydrocarbons Based on Alternate Electron Acceptors (other than molecular oxygen)*; and Section 9, *Natural Bioremediation of Hydrocarbon-Contaminated Ground Water*.

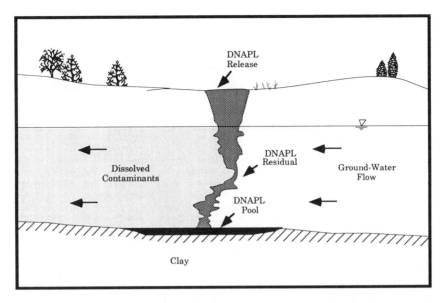

Figure 1.2.   Migration of DNAPL through the vadose zone to an impermeable boundary in relatively homogenous subsurface materials (Huling and Weaver, 1991).

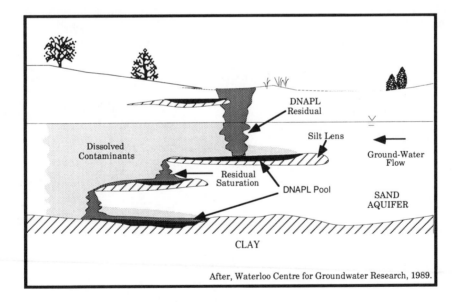

Figure 1.3.   Perched and deep DNAPL reservoirs from migration through heterogeneous subsurface materials (Huling and Weaver, 1991).

Chlorinated solvents are more difficult to bioremediate than petroleum hydrocarbons, and bioremediation efforts are still in the research and development stage. Biological processes for most chlorinated compounds, whether aerobic or anaerobic, require the presence of a primary substrate for cometabolism. Both enhanced and natural bioremediation of chlorinated compounds are widely investigated in laboratory, pilot-scale, and field-scale studies. Results of these efforts are presented in the following sections: Section 5, *Ground-water Treatment for Chlorinated Solvents*; Section 6, *Bioventing of Chlorinated Solvents for Ground-water Cleanup through Bioremediation*; Section 8, *Bioremediation of Chlorinated Solvents Using Alternate Electron Acceptors*; and Section 10, *Natural Bioremediation of Chlorinated Solvents*.

The introduction of microorganisms to the subsurface for bioremediation purposes is discussed in Section 11, *Introduced Organisms for Subsurface Bioremediation*. Although not considered a successful technique at this time due to concerns about survivability of introduced microorganisms, this method may someday be useful at sites sterilized by contamination.

Discussed in each section are basic biological and nonbiological processes affecting the fate of the compounds of interest, documented field experience, performance, repositories of expertise, primary knowledge gaps and research opportunities, favorable and unfavorable site conditions, regulatory acceptance, special requirements for site characterization, and problems encountered with the technology. Although the focus of this paper is bioremediation, remediation of most sites will require use of other technologies not discussed, such as pump and treat, soil washing, etc. The place of bioremediation in the cleanup of hazardous waste sites is still evolving, and evaluation of its effectiveness is under investigation by regulators, researchers, and remediation firms.

REFERENCES

Hinchee, R.E., and H.J. Reisinger.   1987.   A practical application of multiphase transport theory to ground water contamination problems. *Ground Water Monitoring Review*. 7(1):84-92.

Huling, S.G., and J.W. Weaver. 1991. *Dense Nonaqueous Phase Liquids*. Ground Water Issue Paper.   EPA/540/4-91-002.   Robert S. Kerr Environmental Research Laboratory. Ada, Oklahoma.

SECTION 2

IN-SITU BIOREMEDIATION OF SOILS AND GROUND WATER CONTAMINATED WITH PETROLEUM HYDROCARBONS

Robert D. Norris
Eckenfelder, Inc.
227 French Landing Drive
Nashville, Tennessee 37228
Telephone: (615)255-2288
Fax: (615)256-8332

2.1.   INTRODUCTION

This chapter discusses the use of in-situ bioremediation processes to treat ground water and aquifer solids contaminated with petroleum hydrocarbons under aerobic conditions using indigenous microorganisms. Natural bioremediation, the use of nitrate as an electron acceptor, introduced organisms, air sparging to provide oxygen, and treatment of the unsaturated zone are all addressed in other sections. Still other sections address the same topics for chlorinated solvents. Discussions of issues covered in the other sections are included in this document only to the extent that is necessary to adequately address some topics.

2.2.   FUNDAMENTAL PRINCIPLES

Bioremediation, whether of excavated soils, aquifer solids or unsaturated subsurface materials, is the use of microorganisms to convert harmful chemical compounds to less harmful chemical compounds in order to effect remediation of a site or a portion of a site. The microorganisms are generally bacteria but can be fungi. Indigenous bacteria that can degrade a variety of organic compounds are present in nearly all subsurface materials. The use of introduced microorganisms, as discussed in Section 11, has not been shown to be of significant benefit. Microorganisms require certain minerals, usually referred to as nutrients, and an electron acceptor. While a variety of minerals such as iron, magnesium, and sulfur are required by the microorganisms, it is usually only necessary to add nitrogen and phosphorus sources. The other minerals are needed in trace amounts, and adequate amounts are normally found in most ground waters and subsurface materials. The most common electron acceptor used in commercial bioremediation processes is oxygen. Other electron acceptors such as nitrate can be used for some contaminants such as most aromatic hydrocarbons, although restrictions may apply to the levels of nitrate that may be introduced to ground water. This topic is covered in Sections 7 and 8.

Bioremediation systems supply nitrogen, phosphorus, and/or oxygen to bacteria that are present in the contaminated aquifer solids and ground water. Stoichiometrically, it would take approximately three pounds of oxygen to convert one pound of hydrocarbon to carbon dioxide and water. Experience with wastewater treatment indicates that the expressed oxygen requirements are usually near half of the stoichiometric amount. Conversely, some of the oxygen introduced for biooxidation of the contaminants may be consumed by other reactions or is lost through inefficient distribution. As a result, first approximations of oxygen requirements are typically based on the three-to-one ratio.

Nutrient requirements are less easily predicted. If all of the hydrocarbon mass were converted to cell material, the nutrient requirements based on the mass of hydrocarbon to be consumed would be approximated by a ratio of carbon to nitrogen to phosphorous of 100:10:1. The nutrient requirement will be less than this ratio to the extent that direct conversion of hydrocarbons to carbon dioxide and water occurs. Nutrients already exist in the subsurface materials and ground water, nitrogen is fixed by the indigenous bacteria, and nutrients are recycled from dead bacteria. However, adsorption of nutrients by geologic materials can substantially increase the amount of nutrients that have to be introduced in order to distribute nutrients across the contaminated zone. Adsorption may be modest in clean sands but may consume most of the nutrients in silts and clays, especially if the solids have a high natural organic content.

In some instances, nitrogen, phosphorus, or oxygen may be present in sufficient quantities to support degradation of the constituents of interest. In those cases, only one or two of the three elements would need to be added. This is most likely to be the case where low levels of contamination are present. In such cases there may be sufficient nitrogen and phosphorus sources present and only oxygen needs to be provided.

In-situ bioremediation systems for aquifers typically consist of a combination of injection wells (or galleries or trenches) and one or more recovery wells as shown in Figure 2.1. In most instances, the recovered ground water will be treated prior to amendment with nutrients and/or an oxygen source and reinjection. Ground-water treatment has frequently consisted of an air stripper tower or activated carbon but may incorporate an oil/water separator, a biological treatment unit, an advanced oxidation unit, or combinations of treatment units. Treatment of the ground water is likely to be necessary based on regulatory considerations and is beneficial from a process economics perspective when the recovered ground water contains more than a few ppm of biodegradable substances. When the recovered ground water contains low levels of readily degradable constituents, the biodegradable constituents will be degraded within a short distance of the injection point and will not add significantly to the oxygen and nutrient requirements.

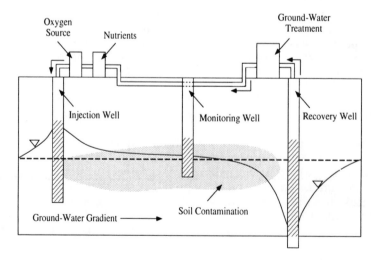

Figure 2.1. Bioremediation in the saturated zone.

Critical to the design of an in-situ bioremediation system are the ground-water flow rate and flow path. Ground-water flow must be sufficient to deliver the required amounts of nutrients and oxygen in a reasonable time frame. The amended ground water should sweep the entire area requiring treatment, and the recovery wells should capture the injected ground water to prevent migration outside the designated treatment zone. In order to ensure that adequate control can be maintained over the ground water, usually only a portion of the recovered ground water is reinjected. The other portion is then discharged by an acceptable method.

In-situ bioremediation systems can be integrated with other remediation technologies either sequentially or simultaneously (Norris et al., 1990). If free phase hydrocarbons are present, a free phase recovery system such as a dual phase pump or skimmer should be used to reduce the mass of the free phase hydrocarbons prior to implementation of bioremediation. In-situ vapor stripping (ISVS) systems (U.S. EPA, 1991a) can serve to both physically remove volatile hydrocarbons and to provide oxygen for biodegradation. ISVS systems can remove residual free phase hydrocarbons as well as constituents adsorbed to both unsaturated materials and aquifer solids exposed during periods of lower water table levels. Depending on the air flow and nutrient availability, hydrocarbons in subsurface solids (including the capillary fringe just above the water table) will undergo biooxidation. The combined mechanisms can serve to reduce significantly the mass of hydrocarbons. As a result, the time and cost of providing nutrients and oxygen through injection of amended ground water may be substantially reduced.

2.3. HISTORICAL PERSPECTIVE

The use of biooxidation for environmental purposes has been practiced for many years. Biological processes have been used to treat wastewater for nearly sixty years (Eckenfelder, 1967). Activated sludge and suspended growth systems have become commonplace for waste treatment in many industries and for municipal waste. This use of biological degradation of organic compounds led to the generation of a wide body of information regarding biodegradability of specific compounds and classes of chemicals, nutrient and electron acceptor requirements, and oxidation mechanisms. Land treatment processes for wastewater, refinery, and municipal wastes have also been practiced for several decades and have generated additional information on nutrient requirements, degradation rates, and other critical parameters affecting biological oxidation (Overcash and Pal, 1979).

In the 1970s, several studies sponsored by the American Petroleum Institute were conducted using the method developed by Richard L. Raymond, Sr., then at Sun Tech., to biologically degrade hydrocarbons in aquifers (Bauman, 1991). This method involved the recovery of ground water with treatment using an air stripper tower and subsequent reinjection following amendment with nitrogen and phosphorus sources (ammonium chloride and sodium orthophosphate salts). Oxygen was generally provided by sparging air at the bottom of the injection well (Raymond et al., 1976). Many of these early tests were conducted prior to the enactment of state mandated cleanup levels. As a result, these tests demonstrated that in-situ bioremediation could reduce the levels of petroleum hydrocarbons in an aquifer, but did not generate documentation of the ability to reach the ground-water quality standards that are necessary in today's regulatory environment.

It was soon recognized that the rate at which oxygen could be introduced by sparging air in a ground-water injection well would limit the effectiveness of the technology. Hydrogen peroxide was identified as a potential method of introducing oxygen (Brown et al., 1984). The solubility in water limits the amount of dissolved oxygen that can be delivered from air to 8 to 10 ppm, unless injection occurs substantially below the water table. Use of pure oxygen in place of air can increase the rate of introduction of oxygen fivefold. Hydrogen peroxide, which decomposes to oxygen and water, is completely soluble in water. Practical considerations, including toxicity towards bacteria, limit hydrogen peroxide concentrations to 100 to 1,000 ppm. Hydrogen peroxide could thus theoretically provide oxygen at 5 to 50 times faster than could sparging air in the injection wells and should result in shorter remediation times. However, the efficiency of delivering oxygen by this method has been quite variable even when favorable results were obtained from laboratory screening tests (Lawes, 1991; Huling et al., 1990; Hinchee and Downey, 1988; and Flathman et al., 1991). Further, as microbial populations decrease as a function of decreasing food source (the contaminants), tolerance toward hydrogen peroxide may also decrease. As a result, hydrogen peroxide may not be the most appropriate oxygen source for many sites.

More recently, many practitioners have used ground-water sparging techniques to introduce oxygen, as discussed in Section 4. In this method, wells or drive points are screened over a narrow interval several feet below the water table. Air is forced into the aquifer under pressure resulting in saturation of the ground water in the vicinity of the injection point. This procedure can also strip volatiles from the ground water and can cause increased rates of migration. Generally, it should be used in conjunction with ground-water capture and/or in-situ vapor stripping systems.

Other approaches to providing an electron acceptor include the use of surfactants to create microbubbles (Michelsen et al., 1990), on site generation of oxygen (Prosen et al., 1991), or use of alternate electron acceptors such as nitrate, as discussed in Sections 7 and 8.

During the time when technology to deliver oxygen was evolving, nutrient sources were being developed, and an understanding of hydrogeological considerations was evolving. Initially, the salts blend developed by Richard L. Raymond, Sr. was used (approximately equal amounts of ammonium chloride and sodium orthophosphate) (Raymond et al., 1978). Some practitioners have changed to potassium salts to reduce the potential for swelling of clays and to tripolyphosphate which will solubilize rather than precipitate iron, calcium, and magnesium (Brown and Norris, 1988).

The need for detailed understanding and control of the site hydrogeology has long been recognized. Many early designs were developed with limited aquifer hydrology test data, and well locations were determined using logic or limited calculations to predict the areas of influence of injection and recovery wells. It has become more common to conduct aquifer tests and use computer models (analytical models suffice for most smaller sites) to locate injection and recovery wells (Falatico and Norris, 1990). This approach can be used to predict remediation times based on oxygen and nutrient demands estimated from contaminant concentrations and ground-water recirculation rates. Models can also be used to evaluate the feasibility of bioremediation at a particular site (Rifai and Bedient, 1987). For readily degradable substances, modeling efforts are more beneficial than laboratory treatability studies and may be less costly. More sophisticated models that also address contaminant and nutrient transport and oxygen uptake are also available (Borden, 1991).

In the mid-1980s, there were few companies with experience in bioremediation of aquifers or soils. Since that time many companies have utilized bioremediation technologies, although claims of experience are frequently overstated. In the last few years this technology has gained the support of the U.S. EPA, as evidenced by the many supportive public statements made by Administrator Reilly, the support of many of the U.S. EPA laboratories, and the creation of U.S. EPA sponsored research, committees, and seminars that have promoted the use of bioremediation.

2.4.   REPOSITORIES OF EXPERTISE

There now exists a large number of organizations and people who are knowledgeable about in-situ bioremediation.   However, many important practical findings have not been adequately shared. Several environmental companies have staff who are experienced and/or knowledgeable in the application of this technology. Some companies have tended to specialize in aboveground systems and may have limited experience in remediation of aquifers. Many large corporations, especially the major oil and chemical companies, have also developed in-house expertise. Several U.S. EPA laboratories (e.g., Robert S. Kerr Environmental Research Laboratory, Ada, Oklahoma), DOD groups (e.g., U.S. Air Force) and universities (e.g., Rice University) have conducted laboratory and, in some cases, field studies on in-situ bioremediation. Typically, some environmental companies, some large site owners, and some EPA laboratories have more experience with the actual application of the technology, while some other groups have more in-depth understanding of the science involved.

2.5.   GENERAL DESIGNS

The most common design is a system that uses a combination of injection and recovery wells, as shown in Figure 2.1. Recovered ground water is treated, typically using an air stripper tower, amended with nutrients, and reinjected. Oxygen is supplied using air sparging in the injection well or by introduction of hydrogen peroxide.

Amended ground water can also be introduced through injection galleries or trenches. This approach is most likely to be used in shallow aquifers.

The above systems can be modified to introduce oxygen by using air spargers located directly within the aquifer, as discussed in Section 5, either in combination with in-situ vapor stripping or using the unsaturated zone as a biofilter.

For shallow aquifers with sandy material, nutrients can be introduced from the surface, allowing percolation of rain water or added water to carry the nutrients into the aquifer. If oxygen is introduced by air sparging, ground-water recovery systems are only required to prevent migration of contaminated ground water.

The design of the ground-water recirculation system is best done using a ground-water flow model (Falatico and Norris, 1990). For most sites, a two-dimensional analytical flow model will be sufficient. The model will allow several design concepts to be evaluated and the most favorable selected. These models can be more effective than laboratory treatability studies to determine feasibility and can be used to make midcourse modifications to operating conditions.

Operating plans should include maintenance of wells and equipment, monitoring schedules, reporting schedules, and milestones for evaluation of system performance so that modifications in operating procedures can be made and, if necessary, additional wells installed.

Each of the systems should incorporate an appropriate monitoring system. Monitoring wells are required to determine the distribution of nutrients and oxygen, and to monitor pH and other ground-water chemistry parameters, ground-water elevations, and changes in contaminant concentrations.

2.6.   LABORATORY TESTING

Laboratory tests can be used as screening tests to determine site feasibility, as treatability tests to determine the rate and extent of biodegradation that might be attained during remediation, and as engineering tests to provide design criteria (U.S. EPA, 1991b).

Screening tests include pH and plate counts to determine if existing conditions are favorable to microbial growth. Respirometer tests, which measure oxygen uptake but do not normally measure disappearance of the contaminant(s), provide confirmation that the microbial population is metabolically active. These tests can be run under a number of nutrient conditions to provide an indication of nutrient effects.

Treatability studies are generally conducted with soil/ground-water slurries. Several conditions are usually tested including unmodified microcosms, nutrient amended microcosms, and biologically inhibited conditions. These tests can measure the rate of change of the constituents of concern as well as changes in pH and microbial populations. The tests provide data on the rate and extent of conversion of contaminants. During bioremediation of hydrocarbons in aquifers, the rate of degradation is usually controlled by the rate of supply of nutrient and oxygen. Under these conditions, laboratory rate data do not extrapolate directly to the field. However, the laboratory data on the rate and extent of removals of hazardous constituents are important for the heavier hydrocarbons, such as heavy crude oil, bunker oil, or coal gas tars. Removal of compounds from these materials is often limited by the reaction kinetics of the microorganisms rather than the rate of supply of some essential nutrient. The extent of biodegradation of oily phase hydrocarbons to microbial biomass or metabolic end products is very site specific.

2.7.   CONTAMINATION LIMITS

The range of contaminant concentrations that are amenable to bioremediation depends on a number of factors. The distribution of contamination may allow remedy through in-situ bioremediation alone. However, if contamination is distributed both above and below the water table, it may be more practical to use other remediation technologies or to combine in-situ bioremediation with other technologies such as free phase recovery, ground-water sparging, and in-situ vapor stripping.

As a class, petroleum hydrocarbons are biodegradable (Gibson, 1984). The lighter, more soluble members are generally biodegraded more rapidly and to lower residual levels than are the heavier, less soluble members. Thus monoaromatic compounds such as benzene, toluene, ethylbenzene, and the xylenes are more rapidly degraded than the two-ring compounds such naphthalene, which are in turn more easily degraded than the three-, four- and five-ring compounds. The same is true for aliphatic compounds where the smaller compounds are more readily degraded than the larger compounds. Branched hydrocarbons degrade more slowly than the corresponding straight-chain hydrocarbons.

Typically, site remediation is concerned with commercial blends of petroleum hydrocarbons. As for individual compounds, the lighter blends are more readily degraded than the heavier blends. For example, gasoline can be biodegraded to low levels under many conditions. Heavier products such as number 6 fuel oil or coal tar, however, contain many higher molecular weight compounds such as five-ring aromatic compounds. These mixtures degrade much more slowly than gasoline and, as a result, significantly lower rates and extents of biodegradation should be anticipated.

Polyaromatic hydrocarbons are present in heavier petroleum hydrocarbon blends and particularly in coal tars, wood treating chemicals, and refinery wastes. These compounds have only limited solubility in water, adsorb strongly to subsurface materials, and degrade at rates much slower than monoaromatic hydrocarbons and most aliphatic and alicyclic compounds found in refined petroleum hydrocarbon products. Because of their low solubility and strong adsorption to solids, their availability for degradation is often the limiting factor in treatment (Brubaker, 1991). They are more likely to be biodegraded in mixtures with more soluble and thus more readily degradable hydrocarbons because the more readily degradable species will support a larger microbial population (McKenna and Heath, 1976).

Because petroleum hydrocarbons are frequently found in the presence of other organic constituents, it is necessary to consider the degradability of other classes of compounds. Nonchlorinated solvents used in a variety of industries are generally biodegradable. Alcohols, ketones, esters, carboxylic acids and esters, particularly the lower molecular weight analogs, are readily biodegradable, but may be toxic at high concentrations due to their high water solubilities. Lightly chlorinated compounds such as chlorobenzene (U.S. EPA, 1986), dichlorobenzene, chlorinated phenols and the lightly chlorinated PCBs are typically biodegradable under aerobic conditions. The more highly chlorinated analogs are more recalcitrant to aerobic degradation but are more susceptible to degradation under anaerobic conditions.

Several of the common chlorinated solvents (chlorinated ethanes and ethenes) can be degraded under aerobic conditions, as discussed in Section 5. This requires the addition of a cometabolite unless certain chemical species such as toluene or phenol are present with the chlorinated species. It is reasonable to expect that some aerobic biodegradation of chlorinated solvents will occur in the presence of petroleum hydrocarbon blends, particularly those containing appreciable amounts of toluene. This is, however, a very site-specific phenomenon and one for which there is not enough documentation to make reliable predictions. Further, many chlorinated solvents can inhibit biodegradation of petroleum hydrocarbons if the solvent species is present at high enough concentrations. Data on biodegradability and other properties of environmental interest are available from several handbooks (Montgomery, 1991; Montgomery and Wilkom, 1990; Howard, 1989 and 1990; and Verschueren, 1983).

Petroleum hydrocarbons can generally be mineralized; i.e., converted to carbon dioxide and water. The extent of conversion that is likely to occur is greatest for the lighter molecular weight constituents. For gasoline, the extent of conversion is largely limited by the efficiency and completeness of the distribution of nutrients and an electron acceptor. For the heavier petroleum hydrocarbons, especially polynuclear aromatic hydrocarbons (PAHs), the limiting factor may be the rate of solubilization, the release from interstitial pore spaces, or the rate of degradation of the higher molecular weight constituents.

Concentrations of contaminants that are toxic or large quantities of oily phase material that do not permit penetration of nutrients and/or an electron acceptor are not appropriate for in-situ bioremediation. Toxicity seldom occurs where petroleum hydrocarbons are the only contaminants. Toxicity can occur with some chlorinated solvents and with very soluble compounds such as alcohols. The levels at which a specific compound is toxic will be to some extent site specific as the microbial communities have substantial capacity to adapt. Approximations of concentrations at which specific compounds are toxic can be obtained from several handbooks (Montgomery, 1991; Montgomery and Wilkom, 1990; Howard, 1989 and 1990; and Verschueren, 1983).

Bioremediation is not generally applicable to metals but may incidentally mobilize or immobilize various metals. Generally, the presence of metals has little direct effect on the bioremediation process (Robert Norris, personal experience). While some metals such as mercury can be toxic to bacteria, the microbial population frequently adapts to the concentrations present. The effect of metals or organics on the microbial population of a specific site can be tested through plating experiments or treatability tests.

When viscous materials such as the heavier fuel oil blends are present at concentrations that prevent the flow of water and diffusion of nutrients and electron acceptors, bioremediation will be impractical. The concentration at which this occurs will vary with soil type but generally will be above 20,000 mg/kg.

Very high concentrations of contaminants will create very high oxygen and/or nutrient demands. Meeting these demands might require excessively longer times and higher costs than other technologies (Piontek and Simpkin, 1992). High levels of contamination are a bigger problem with low or marginally permeable aquifers than with highly permeable aquifers. In some cases, provision of oxygen through ground-water sparging may provide a suitable approach to higher levels of contamination.

2.8.   SITE CHARACTERIZATION

Two important aspects of site characterization frequently receive less attention than they should. While implementation of this or any on-site or in-situ technology requires delineation of the extent of contamination, including the presence and extent of oily phase material, the concentrations of contaminants on aquifer solids is often overlooked. The solubility of petroleum hydrocarbons is low and thus the preponderance of the hydrocarbon mass is associated with the solids and not in the dissolved phase. For sites contaminated with petroleum hydrocarbons, quantities associated with aquifer solids are far more important than ground-water concentrations. Even when numerous samples of both cores and ground waters have been analyzed by currently available standard analytical methods, the total mass of hydrocarbons may not be accurately determined. The total mass calculated from component specific analyses for volatiles, semivolatiles, polynuclear aromatics, base neutrals, and acid extractables do not account for the total mass. Nonspecific analysis such as Total Petroleum Hydrocarbon (TPH) analyses can measure components that are not of interest; e.g., asphalt particles, do not measure the most volatile compounds, and can yield highly variable results as shown in studies where split samples have been sent to different laboratories (Anonymous, 1992).

Even without analytical considerations, obtaining representative data is sometimes difficult, particularly for sites with heterogeneous conditions and/or multiple sources. Even with extensive sampling, it is quite likely that the total mass of contaminants at a specific site will not be known within 50 percent; however, if analyses of the aquifer solids are not conducted, the uncertainty can be an order of magnitude or more.

The second important aspect of site characterization that is frequently slighted is site hydrogeology. Because the rate of remediation of petroleum hydrocarbons in saturated materials is almost always controlled by the rate of distribution of the nutrients and oxygen source, and thus the rate of ground-water recirculation, aquifer hydrogeological properties are critical. For easily biodegraded materials such as gasoline, it is more important to model the ground-water flow than it is to conduct laboratory treatability studies. Site characterizations should include aquifer tests such as 24-hour pump tests. Relatively simple analytical flow models can provide a good approximation of ground-water recovery and injection capabilities and thus the feasibility of providing nutrients and oxygen in an acceptable time frame.

It is necessary to identify nearby ground-water receptors in order to design a capture system for the injected water that protects adjacent ground-water supplies and to be able to evaluate the potential impact of residual nutrients that may discharge to surface water.

The concentrations of other potential contaminants that are not biodegradable, such as heavy metals, should be determined because bioremediation processes are not likely to appreciably change the concentration of these species. If they are present above regulated levels, it may be necessary to combine bioremediation with another technology or select another remediation strategy.

Microbial populations offer an indication of whether site conditions will support microbial activity that will degrade petroleum hydrocarbons. Tests can be made for heterotrophic (total) microbial populations or for bacteria that can utilize the contaminant of interest. This can be a useful tool to screen for conditions where bacteria have been negatively impacted by the site conditions. Although failure of soils or aquifer solids to contain a viable microbial community capable of degrading a range of petroleum hydrocarbons is rare (<<1% of sites), early identification of such a problem is important.

2.9.   FAVORABLE SITE CONDITIONS

2.9.1. Solubility

The more soluble hydrocarbons are readily biodegraded and can be partially captured by recovery wells for surface treatment. Petroleum hydrocarbons are not sufficiently soluble to be treated by pump and treat alone.

2.9.2. Volatility

Volatility does not affect biodegradation; however, the volatility of the contaminant does determine if it can be treated by ground-water sparging combined with biodegradation in the unsaturated zone, or with in-situ vapor stripping, or in-situ vapor stripping combined with dewatering.

2.9.3. Viscosity

Highly viscous hydrocarbons are not as easily biodegraded because it is difficult to establish contact among contaminant, bacteria, nutrients, and an electron acceptor.

2.9.4. Toxicity

Contaminants may be toxic or inhibitory to the microbial community. Frequently, the bacteria have adapted to the presence of these compounds. This can usually be readily determined by performing plate counts of subsurface materials and ground-water samples or conducting treatability tests to determine the effect of potential toxicants on the rate and extent of biodegradation.

2.9.5. Permeability of Soils and Subsurface Materials

The greater the permeability, the easier it is to distribute nutrients and an electron acceptor to the contaminated solids and ground water. Of course, these conditions also tend to lead to greater extent of contamination. The importance of permeability increases with the mass of contaminant to be addressed and the urgency of completing the remediation process.

2.9.6. Soil Type

In addition to permeability, soil type also impacts the degree of adsorption of contaminants and nutrients by the soils. Sand and gravel are the most favorable soil types for nutrient transport; clays are the least favorable. Karst formations allow for rapid recovery and introduction of amended ground water; however, prediction and control of flow paths may be difficult or severely limited. The soil organic matter content (e.g., humates) impacts the movement of petroleum hydrocarbons through the aquifer.

2.9.7. Depth to Water

Depth to ground water should be considered not so much as a favorable or unfavorable characteristic but as a factor to be taken into consideration in designing a system. The greater the depth to water, the greater head that can be provided at injection points, and thus the greater the potential injection rates that can be obtained. Shallower water tables limit the head that can be attained and are more favorable to the use of injection galleries. Air sparging, when used as an oxygen source, also has the potential to transfer volatiles to the unsaturated zone and thus the surface air. Efficient capturing of these gases requires an adequate unsaturated interval if an in-situ vapor stripping system is used or if the unsaturated materials are used as a biofilter. Significant depths to water can add to the cost of installation, but will also add to the cost of other alternatives as well.

2.9.8. Mineral Content

Calcium, magnesium, and iron can cause precipitation of nutrients and caking in water lines. This can be minimized by using tripolyphosphates, which sequester these minerals. However, tripolyphosphate will form precipitates with these minerals unless present in amounts equal to or greater than a 1:1 molar ratio.

2.9.9. Oxidation/Reduction Potential

Iron can also be a problem because natural biooxidation of petroleum hydrocarbons can consume nearly all of the available oxygen in the ground water. As a result of these reduced conditions, ferric iron can serve as an electron acceptor for anaerobic degradation of some hydrocarbons. In this process, ferric iron is reduced to ferrous iron, which is more soluble. When oxygenated ground water is introduced into the formation, the less soluble oxidized form of iron (Fe^{+3}) will form and precipitate. This can reduce the permeability of the formation. (However, if the aquifer is maintained in the oxidized state, further dissolution of iron should not occur. In The Netherlands, ground water in the vicinity of production wells is routinely oxygenated to reduce the amount of iron in the production well water.)

2.9.10. pH

Bioremediation is favored by near neutral pH values (6 to 8). However, in aquifers where natural pH values are outside this range, biodegradation may proceed without hindrance. Biodegradation appears to proceed quite well, for instance, in the New Jersey Pine Barrens where pH values of 4.5 to 5 are common (Brown et al., 1991). Where the pH has been shifted away from neutral by manmade changes, biodegradability is likely to be impaired (Robert Norris, personal experience).

2.10. INFRASTRUCTURE AND INSTITUTIONAL ISSUES

One advantage of in-situ treatment systems is the ability to install and conduct remediation with minimal disruption to the site. Implementation does, however, require the installation of wells, transfer lines, aboveground systems for amending the injection water with nutrients and an electron acceptor source and, if necessary, a treatment system for removal or reduction of the contaminant in the recovered ground water prior to reinjection and/or discharge to an alternate receptor. The size of the treatment area will depend on the ground-water flow and the treatment design. In most cases the size and appearance of the aboveground treatment infrastructure are acceptable.

The acceptance of in-situ bioremediation as a remediation technology by the public and various regulatory agencies has been generally favorable over the last seven or eight years and has improved significantly over the last two or three years with the support of the U.S. EPA, many state agencies, as well as favorable publicity in trade journals and the popular press. As an in-situ technology that is viewed as a natural process that results in destruction rather than relocation of the contamination, in-situ bioremediation meets many of the objectives of State and Federal agencies. Questions of efficacy (biodegradability) and production of toxic intermediates have infrequently been an issue with the treatment of petroleum hydrocarbons.

Issues tend to be mostly site-specific. One frequent issue is the discharge of treated or untreated ground water. In many instances, several permits are required. Work has been delayed because a particular permit was delayed or was altogether unobtainable. In order to maintain hydraulic control over the aquifer, it is generally necessary to reinject only a portion of the recovered ground water. Some states regulate reinjection wells and galleries as Class V wells. The remaining water can be discharged to a municipal sewer. Where sewer or water treatment systems are near or exceed capacity, sewage discharge permits may not be obtainable.

If the remaining ground water is discharged to surface water, a National Pollutant Discharge Elimination System (NPDES) permit is usually required. States in arid regions usually require a special permit for extraction of ground water. While site remediations conducted under Superfund allow work to proceed without formally obtaining state and local permits, the standards and requirements of the permits must be met.

2.11.   PERFORMANCE

The ability to meet relevant regulatory end points depends both on the end points and the limits of the technology. End points can be State mandated levels, risk based levels, Federal mandated levels, or Toxic Characteristic Leaching Potential (TCLP) based levels. The targeted end points can vary significantly, and the specific levels set for a given site often determine whether end points will be met by the specific technology employed. Particularly troublesome are State regulations that set levels at or below the detection limit or at background. Because it is a nonspecific analysis, using background TPH levels as the remediation goal can create difficulties in interpreting data and lead to misleading conclusions regarding the performance of the system.

Under ideal conditions, in-situ bioremediation can reduce petroleum hydrocarbon levels to nondetectable levels (10 mg/kg). This is more easily obtained with the lighter blends in permeable and homogeneous formations where placement of injection and recovery wells (galleries, etc.) is unencumbered. Generally, for lighter petroleum blends, the hardest regulatory end point to meet is the benzene limit. Although benzene is highly biodegradable, MCLs for benzene are at least an order of magnitude lower than for other specific light hydrocarbon constituents. As a result, if the benzene end point can be reached, the level for the other components will most probably be met as well.

For heavier petroleum hydrocarbons, BTEX compounds (benzene, toluene, ethylbenzene, and xylenes) may not be present in significant concentrations to be of concern. Typically, TPH will be the target analysis to be met. The heavier the petroleum mixture, the more probable there will be residuals of very slowly degraded components. These components tend to have low water solubilities, which can limit their rate of degradation. If TPH is the only criterion, the measurements will not determine which petroleum hydrocarbon components have gone untreated. Compounds that are not of environmental concern may contribute to reported TPH values and thus complicate interpretation.

Polyaromatic hydrocarbons can be difficult to treat to the regulated levels. The MCLs for many of these compounds are low because they are suspected carcinogens. The rate of release of PAHs from subsurface solids may be too slow to support an active microbial population and degradation rates may be impractically slow. Fortunately, the degradability of these compounds is better in mixtures containing lower molecular weight compounds found in many commercial petroleum products. Available data on the limits of PAH degradation under in-situ bioremediation conditions are limited and contradictory, and thus predictions of treatment limits are likely to be unreliable.

2.12.   PROBLEMS

Inadequate characterization of a site can result in a bioremediation system being underdesigned. If the total mass of contamination is underestimated, a specific design will take longer to achieve the remediation goals than predicted from the available data. If the site hydrogeology is not adequately characterized, the production rate of recovery wells or, more likely, the rates at which injection wells can receive water may be overestimated. If this latter situation occurs, it will take longer to provide the required nutrients and electron acceptor. Since provision of nutrients and/or oxygen is frequently the rate controlling step, the remediation may take proportionately longer.

The capacity of recovery wells, and particularly injection wells, tends to decrease with time. The deterioration of injection wells can result from movement of fines, precipitation of minerals, or from excessive microbial growth in or in the immediate vicinity of the screened interval of the injection well. Proper selection of a gravel pack and installation and development of the wells will reduce the propensity for problems.

Mineral clogging of the well screen and formation can occur because the chemistry of the injection water is different from that of the ground water. Typically, ground water in an aquifer contaminated with petroleum hydrocarbons will have a low oxidation potential because natural biodegradation will have utilized most of the dissolved oxygen. Frequently this results in elevated dissolved minerals, especially iron. Recovery and treatment of ground water typically introduces oxygen into the water even if an oxygen source is not added. Reinjection of this water can result in precipitation of iron and other metals when the injection water mixes with the ground water.

In some instances it may be preferable to discharge all of the recovered ground water and reinject clean water from another source. Other sources of water that have been used are city water and uncontaminated ground water from another part of the same or adjacent aquifer.

Calcium, magnesium, and iron will form precipitates with orthophosphates when orthophosphate salts are used as nutrients. The formation of precipitates should be evaluated with laboratory tests during the design phase. The use of adequate levels of tripolyphosphate salts can alleviate precipitation problems.

Reduced aquifer permeability can also result from swelling of clays if sodium salts of phosphates are used as the nutrient source. In such materials potassium salts should be used.

Biological growth can reduce permeability and/or restrict flow through well screens. This can be addressed by periodically adding higher levels of hydrogen peroxide and surging the wells.

Use of dilute hydrochloric acid to clean the wells may also work, particularly when mineral deposits are the primary problem (Driscoll, 1986). For treatment of excessive microbial growth, however, hydrogen peroxide has the advantage that the dead microbial mass is in the form of particles as opposed to the slimy material that can form following acidification. Removal of the biological mass will be facilitated by a more flocculent mass as opposed to a slimy mass.

It has been suggested that the addition of nutrients in high concentration batches instead of continuous addition at low concentration might reduce the tendency for microbial growth in the well bore and the immediate vicinity of the injection point.

After the system has been in operation for an extended period, it may become apparent that the distribution of nutrients and oxygen is not as anticipated. Frequently this can be corrected by adjusting the relative rates of ground-water recovery or reinjection in the various wells. These adjustments are more efficiently made using a ground-water flow model. Determination of ground-water elevations in monitoring wells will determine within a few days whether or not the adjustment in flows is having the desired effect. Statistically significant changes in contaminant, nutrient, or dissolved oxygen levels are likely to take several weeks to a few months to be observed. In some instances, adjusting the flows between wells may not produce the desired effect. It may then be necessary to add an additional well(s).

Identifying problems with nutrient distribution and thus impact on contaminant levels in a timely fashion requires that monitoring wells be properly located. Wells need to be located so that flows in different directions can be determined and at distances that produce changes in water chemistry within a reasonable time frame. The distance between injection wells and the nearest monitoring wells should be based on predicted flow times rather than distance. The time of travel for a conservative tracer between the injection well and the nearest monitoring well should be on the order of one week.

If either nutrients or contaminants appear in monitoring wells that are outside the treatment zone, it may be necessary to change the relative flows in the injection and recovery wells. In particular, it may be necessary to reduce the fraction of the recovered ground water that is being reinjected.

Some state regulations (e.g., New Jersey) require that the final nutrient constituent levels in the ground water be at or below background levels at the completion of the project. In order to avoid problems meeting this requirement, it is necessary to use the minimum amount of nutrients that are needed to complete biodegradation across the site. Since nutrient requirements are a function of many factors, it is difficult to determine the total amount required *a priori*. It is necessary to monitor nutrient distribution across the site during remediation and adjust nutrient addition rates to balance requirements of the bacteria, nonbiological removal in the aquifer, and the nutrient concentrations required to close the site.

2.13. SITE PROPERTIES vs COSTS

In-situ bioremediation costs are dependent on a number of factors including site conditions, remedial goals, the design of the system, and the operating and monitoring schedule.

2.13.1. Mass of Contaminant

The greater the mass of contamination present, the greater the nutrient and electron acceptor requirements. This increases not only the chemical costs, but increases either the time to achieve remediation or requires greater capital expenditure for wells, pumps, and aboveground treatment.

2.13.2. Volume of Contaminated Aquifer

The greater the volume of aquifer solids and ground water subject to treatment, the greater the number of injection and recovery points that will be required, or the time to achieve remediation will be longer.

2.13.3. Aquifer Permeability/Soil Characteristics

For a given size plume and contaminant mass, it will generally be more expensive to remediate a low permeability aquifer than to remediate a more permeable aquifer because either longer remediation times or more injection and recovery points will be required.

2.13.4. Final Remediation Levels

The more stringent the remediation goals, the more costly will be the remediation in most cases. For gasoline and other light petroleum hydrocarbon spills in relatively permeable and homogeneous aquifers, the time to proceed from a less stringent remediation goal to a more stringent remediation goal might not be very long. However, for the heavier hydrocarbon blends, particularly those containing PAHs, the slow dissolution and thus biodegradation of the least soluble components may limit the rate of bioremediation. Long periods of time may be required to meet stringent ground-water quality standards. A small change in the remediation goal could thus make a large change in the time to attain the remediation goal and may also increase the size of the aquifer zone that requires active treatment.

2.13.5. Depth to Water

The depth to water will affect the design of the system as well as the cost. For systems using injection wells, the greater the depth to water, the more head pressure and thus the greater the flow of injected water that can be introduced. This can reduce the number of wells needed or shorten the remediation time. However, the cost of installing wells increases with depth. Very shallow aquifers may be treated using injection galleries, or through percolation of nutrients with air sparging and thus be relatively inexpensive to construct and operate.

2.13.6. Monitoring Requirements

Monitoring costs can be substantial. The number of wells to be monitored, the frequency of monitoring, the number and type of parameters all contribute to the costs. Monitoring should be designed to provide a basis for evaluation of progress, to identify conditions that require process modifications, and to ensure that the system is under hydraulic control. Monitoring data that does not serve as a basis for making decisions should be avoided. Unnecessary monitoring adds to the costs of data acquisition, data interpretation, and report writing and reading. Unnecessary data can also impede interpretation of the critical issues.

2.13.7. Contaminant Properties

The extent to which a compound will be recovered with captured ground water is dependent on its solubility or, more precisely, its octanol/water partition coefficient as well as the organic content of the solids. The larger the proportion of the contaminant mass that is recovered, the less time and expense required to provide sufficient nutrients and electron acceptor. On the other hand, treatment costs for the recovered ground water may increase with the concentration of the contaminant in the recovered water.

2.13.8. Location of Site

The location of a site can also impact the costs. Remote sites will have higher costs of providing labor due to travel and housing costs. This is typically much more important for small sites, particularly for a system whose operations are highly automated and technicians are not required on a daily basis.

The use of air sparging techniques offers the potential to reduce the costs of in-situ bioremediation. The depth to water, type of subsurface material, and saturated interval of the aquifer will all affect the costs. Shallow aquifers beneath sandy materials permit nutrients to be added from the surface. Large, saturated intervals permit large radius of influences of sparging wells and thus smaller numbers of sparge wells. Stratification of subsurface materials also affects the radius of influence. For greater detail on air sparging see Section 4.

2.14. PREVIOUS EXPERIENCE WITH COSTS

Costs for bioremediation are not easily generalized. As previously discussed, many factors affect the cost of remediation. The number of completed and documented in-situ bioremediation projects with readily available cost data is small compared to the variables affecting the costs. Frequently, available cost data, especially from larger sites, includes but does not define costs for many other activities associated with the site remediation.

For the same site conditions and contaminant distribution, the cost of bioremediation can vary significantly depending on the specific design. For instance, incorporating more recovery and injection wells will increase the capital costs but may reduce the operating and maintenance costs by reducing the total time of remediation.

The choice of an oxygen source (or an alternate electron acceptor) may have a large impact on costs. Using hydrogen peroxide instead of oxygen will increase monthly operating costs, but may reduce overall operating costs by shortening the period of operation and thus the time over which operating, monitoring, and reporting costs are incurred. Provision of oxygen through air sparging has the potential to substantially shorten the time of remediation and costs, especially for heavier hydrocarbons. For lighter petroleum hydrocarbons, the reduced time and cost of supplying oxygen may be offset by the additional costs for a system to capture and treat air from the unsaturated zone. Air sparging using the unsaturated zone as a vapor phase biotreatment system could prove to be the lowest cost system for volatile hydrocarbons.

In addition to differences in designs, the contractor can impact costs through the selection of equipment as well as efficiency of construction, permitting, and operation. Generally, the use of good quality components and automated equipment will minimize overall costs.

Limited anecdotal (R.A. Brown, 1992) and personal information indicate that in-situ bioremediation of light petroleum products at leaking underground storage tank sites has cost from one to 1.5 million dollars for 0.5 to one-acre sites and required from less than one to up to five years. Costs per acre would be expected to decrease significantly with scale-up. Two systems that included an in-situ venting system which served to address a large portion of the contamination at the capillary zone, cost 0.7 to one million dollars and took approximately three years. Extrapolation from an air sparging system that treated chlorinated solvents through physical removal suggests costs of 0.3 to 0.5 million dollars and a time frame of one to 1.5 years.

A recent U.S. EPA document (U.S. EPA, 1991c) provides some cost information for ongoing current projects. This cost data may not be inclusive of all costs and may not sort out costs where multiple technologies are being used.

-    New York: Gasoline contamination over approximately one acre with a depth to water of ten feet. System consists of an infiltration trench for nutrients and hydrogen peroxide and three 80 gpm recovery wells. Initiated in January 1989. Costs are reported to be $250K.

-    Iowa: Ground water contaminated with PAHs and BTEX. In-situ system. Construction costs reported as $149K with anticipated additional costs of $1.5M.

-    Kansas: Approximately 700K cubic feet of aquifer contaminated with BTEX. Combined in-situ soil flushing with bioremediation using nitrate. Bioventing is also being considered. Reported expenditures were $275K with anticipated additional costs of $650K.

-   California: Approximately 3,000 cubic yards of aquifer contaminated with diesel and gasoline. System consisted of a closed loop system with hydrogen peroxide as the oxygen source. In-situ vapor stripping and soil flushing were also incorporated. Started November 1988 and completed March 1991. Costs reported as $1.6M.

-   Michigan: An approximately 1/4-acre site was contaminated with gasoline. System consisted of infiltration gallery and injection wells. Air and hydrogen peroxide were used as oxygen sources. Source area treatment lasted approximately 1.5 years. Costs were approximately $600K, of which sampling and analytical costs were a significant portion. Some residual in downgradient edge of plume will be addressed with air sparging.

-   Texas: Approximately 20 acres contaminated with BTEX, some chlorinated solvents, and other organics. In-situ bioremediation is being used to augment pump and treat. Both pure oxygen and nitrate are being used as electron acceptors. Recovered ground water is being treated in an aboveground bioreactor. Ground water from a clean portion of the aquifer is being used for injection. Capital costs were approximately $5M, including the water treatment plant. Upgrading the initial pump-and-treat system to include in-situ bioremediation cost $200K in capital and $150K in pilot testing and engineering.

2.15. REGULATORY ACCEPTANCE

Acceptance of the technology on a specific site is, as for any technology, impacted by those criteria normally used to evaluate technologies:

-   Is it appropriate for the contamination of concern?
-   Is it a permanent remedy?
-   Is it implementable under the specific site conditions?
-   Can the technology reach the site specific remediation goals?
-   Is the technology innovative?
-   Has the technology been demonstrated?
-   Can the technology be implemented without violating the intent of local or state regulations?
-   Will its application be protective of public health during its construction and as a result of its performance?

Specifically favorable to in-situ bioremediation are the benefits of being able to avoid bringing contaminated subsurface materials to the surface, minimization of interruption of ongoing commercial operations, low profile of operations, destruction rather than transport of the contaminants and, frequently, costs.

Areas of concern specific to in-situ bioremediation are: (1) State regulations prohibiting the injection of water not meeting drinking water standards; (2) residual levels of nutrient components; (3) potential for formation of nitrate from ammonium; and (4) concern for maintaining hydraulic control over the contaminated ground water and the injected nutrients.

2.16.   KNOWLEDGE GAPS

Knowledge gaps include both those items that are not understood well by anyone and the myriad of small pieces of information that are known by various individuals and consultants that have not been disseminated in any organized fashion.   Information obtained from one site tends to be generalized for all sites. In some instances conclusions drawn from performance of one design are used to evaluate systems of such different design characteristics that the conclusions are not at all valid. Many knowledge gaps could be considerably narrowed if an efficient exchange of information could be achieved and maintained.

Specific areas where increased understanding would be very beneficial to the implementation of in-situ bioremediation are (Bartha, 1991):

- Identification of the real cause and effects of the difficulties that have been observed on some sites.
- Better correlation between laboratory test results and field performance.
- Greater understanding of nutrient transport.   How do different nutrient sources move through various types of subsurface material and what can be done to facilitate nutrient transport?
- Selection, control, and enhancement of oxygen distribution under a variety of site conditions.
- Greater understanding of conditions under which aquifer permeability will be reduced and how to prevent aquifer blockage.
- Better models of nutrient, oxygen, and contaminant transport and biodegradation rates.
- Better understanding of natural attenuation as an alternative to active remediation or subsequent to active remediation.
- Better cost data on completed projects.
- Methods of addressing low permeability aquifers.
- Methods of solubilizing the higher molecular weight hydrocarbons.
- Increased understanding of the effects and limits set by site conditions.
- Improved methods of engineering and site management.
- Improved methods for estimating contaminant mass, including analytical procedures.
- Better understanding of degradation pathways even though degradation pathways are much better understood for petroleum hydrocarbons than for most other classes of compounds.
- Better assessment protocols.

REFERENCES

Anonymous. 1992. TPH results vary significantly from lab to lab. *The Hazardous Waste Consultant*. January/February. pp. 1.7 - 1.10.

Bartha, R. 1991. Utilizing bioremediation technologies: Difficulties and approaches. *Bioremediation Workshop*. Interdisciplinary Bioremediation Working Group. Rutgers University, New Brunswick, New York.

Bauman, B. 1991. Biodegradation research of the American Petroleum Institute. Presented at: *In Situ Bioreclamation: Application and Investigation for Hydrocarbons and Contaminated Site Remediation*. San Diego, California. March 19-21, 1991.

Borden, R.C. 1991. Simulation of enhanced in situ biorestoration of petroleum hydrocarbons. In: *In Situ Bioreclamation: Application and Investigation for Hydrocarbons and Contaminated Site Remediation*. Eds., R.E. Hinchee and R.F. Olfenbuttel. Butterworth-Heinemann. pp. 529- 534.

Brown, R.A. 1992. Personal communication. Groundwater Technology, Inc. Trenton, New Jersey.

Brown, R.A., Dey, J.C. and McFarland, W.E. 1991. Integrated site remediation combining groundwater treatment, soil vapor extraction, and bioremediation. In: *In Situ Bioreclamation: Application and Investigation for Hydrocarbons and Contaminated Site Remediation*. Eds., R.E. Hinchee and R.F. Olfenbuttel. Butterworth-Heinemann. pp. 444-449.

Brown, R.A., and R.D. Norris. 1988. U.S. Patent 4,727,031. *Nutrients for Stimulating Aerobic Bacteria*.

Brown, R.A., Norris, R.D., and Raymond, R.L. 1984. Oxygen transport in contaminated aquifers. In: *Proceedings of the Petroleum Hydrocarbon and Organic Chemicals in Groundwater: Prevention, Detection, and Restoration*. National Water Well Association. Houston, Texas.

Brubaker, G.R. 1991. In situ bioremediation of PAH-contaminated aquifers. In: *Proceedings of the Petroleum Hydrocarbons and Organic Chemicals in Ground Water: Prevention, Detection, and Restoration*. Houston, Texas. pp. 377-390.

Driscoll, F.G. 1986. *Groundwater and Wells*. Johnson Division. St. Paul, Minnesota. pp. 630-658.

Eckenfelder, W. Wesley, Jr. 1967. *Industrial Water Pollution Control*. McGraw-Hill Book Company. New York, New York.

Falatico, R.J., and Norris, R.D. 1990. The necessity of hydrogeological analysis for successful in situ bioremediation. In: *Proceedings of the Haztech International Pittsburgh Waste Conference*. Pittsburgh, Pennsylvania. October 2-4, 1990.

Flathman, P.E., K.A. Khan, D.M. Barnes, J.H. Caron, S.J. Whitehead, and J.S. Evans. 1991. Laboratory evaluation of hydrogen peroxide for enhanced biological treatment of petroleum hydrocarbon contaminated soil. In: *In Situ*

Bioreclamation: Application and Investigation for Hydrocarbons and Contaminated Site Remediation. Eds., R.E. Hinchee and R.F. Olfenbuttel. Butterworth-Heinemann. pp. 125-142.

Gibson, D.T. 1984. *Microbial Degradation of Organic Compounds*. Microbiology Series. Marcel Dekker, Inc. New York, New York.

Hinchee, R.E., and D.C. Downey. 1988. The role of hydrogen peroxide in enhanced bioreclamation. In: *Proceedings of the Petroleum Hydrocarbons and Organic Chemicals in Ground Water: Prevention, Detection and Restoration*. *Vol 2*. National Water Well Association. pp. 715-721.

Howard, P.H. 1989. *Handbook of Environmental Fate and Exposure Data for Organic Chemicals: Volume I. Large Production and Priority Pollutants*. Lewis Publishers. Chelsa, Michigan.

Howard, P.H. 1990. *Handbook of Environmental Fate And Exposure Data For Organic Chemicals: Volume II. Solvents*. Lewis Publishers. Chelsa, Michigan.

Huling, S.G., B.E. Bledsoe, and M.V. White. 1990. *Enhanced Bioremediation Utilizing Hydrogen Peroxide as a Supplemental Source of Oxygen: A Laboratory and Field Study. Final Report*. NTIS PB90-183435/XAB. EPA/600/2-90/006.

Lawes, B.C. 1991. Soil-induced decomposition of hydrogen peroxide. In: *In Situ Bioreclamation: Application and Investigation for Hydrocarbons and Contaminated Site Remediation*. Eds., R.E. Hinchee and R.F. Olfenbuttel. Butterworth-Heinemann. pp. 143-156.

McKenna, E.J., and R.D. Heath. 1976. *Biodegradation of Polynuclear Aromatic Hydrocarbon Pollutants by Soil and Water Microorganisms*. University of Illinois Research Report No. 113. UILU-WRC-76-0113.

Michelsen, D.L., M. Lofti, and D.L. Violette. 1990. Application of air microbubbles for treatment of contaminated groundwater. In: *Proceedings of the HMCRI - 7th National RCRA/Superfund Conference and Exhibition*. St. Louis, Missouri.

Montgomery, J.H. 1991. *Groundwater Chemical Desk Reference. Vol. II*. Lewis Publishers. New York, New York.

Montgomery, J.H., and L.M. Wilkom. 1990. *Groundwater Chemical Desk Reference. Vol. I*. Lewis Publishers. New York, New York.

Norris, R.D., S.S. Sutherson, and T.J. Callmeyer. 1990. Integrating different technologies to accelerate remediation of multiphase contamination. Presented at *NWWA Focus Eastern Regional Groundwater Conference*. Springfield, Massachusetts. October 17-19, 1990.

Overcash, M.R., and D. Pal. 1979. *Design of Land Treatment Systems for Industrial Wastes*. Ann Arbor Science. Ann Arbor, Michigan.

Piontek, K.R., and T.S. Simpkin. 1992. Factors challenging the practicability of in situ bioremediation at a wood preserving site. In: *Proceedings of the 85th Annual Meeting and Exhibition of the Air and Waste Management Association*. Kansas City, Misssouri. June 21-26, 1992.

Prosen, B.J., W.M. Korreck, and J.M. Armstrong. 1991. Design and preliminary performance results of a full scale bioremediation system utilizing an on-site oxygen generation system. In: *In Situ Bioreclamation: Applications and Investigations for Hydrocarbons and Contaminated Site Remediation.* Eds., R.E. Hinchee and R.F. Olfenbuttel. Butterworth-Heinemann. pp. 523-528.

Raymond, R.L., V.W. Jamison, J.O. Hudson. 1976. *AIChE.* Symposium Series. 73:390-404.

Raymond, R.L., V.W. Jamison, J.O. Hudson, R.E. Mitchell, and V.E. Farmer. 1978. *Field application of subsurface biodegradation of hydrocarbon in sand formation.* Project No. 307-77. American Petroleum Institute, Washington, D.C. 137 pp.

Rifai, H.S., and P.B. Bedient. 1987. Bioplume II, Two-dimensional modeling for hydrocarbon biodegradation and in situ restoration. In: *In Proceedings of the NWWA/API Conference on Petroleum Hydrocarbons and Organic Chemicals in Ground Water: Prevention Detection, and Restoration.* National Water Well Association. Houston, Texas. pp 431-450.

U.S. Environmental Protection Agency. 1986. *Microbiological Decomposition of Chlorinated Aromatic Compounds.* EPA 600/2-86/090.

U.S. Environmental Protection Agency. 1991a. *Soil Vapor Extraction Technology. Reference Handbook.* EPA/540/2-91/003

U.S. Environmental Protection Agency. 1991b. *Guide for Conducting Treatability Studies Under CERCLA: Aerobic Biodegradation Remedy Screening. Interim Guidance.* EPA/540/2-91/013A.

U.S. Environmental Protection Agency. 1991c. EPA/540/2-91/027. *Bioremediation in the Field.*

Verschueren, K. 1983. *Handbook of Environmental Data on Organic Chemicals.* 2nd Edition. Van Nostrand Reinhold. New York, New York.

SECTION 3

BIOVENTING OF PETROLEUM HYDROCARBONS

Robert E. Hinchee
Battelle Memorial Institute
505 King Avenue
Columbus, Ohio 43201-2693
Telephone: (614)424-4698
Fax: (614)424-3667

3.1. FUNDAMENTAL PRINCIPLES

Bioventing is the process of supplying air or oxygen to the unsaturated zone to stimulate aerobic biodegradation of a contaminant. Bioventing is applicable to any contaminant that is biodegradable aerobically. Air can be injected through boreholes screened in the unsaturated zone, or air can be extracted from boreholes, pulling air from the surface into a contaminated area.

For the purposes of this manuscript, the term "bioventing" will be reserved to processes occurring above the water table. The term "air sparging" as discussed in Section 4 is a separate technology designed to treat contamination below the water table. The two technologies are often used in combination. Obviously, once sparged air rises above the water table, it can also biovent the unsaturated zone. This section will focus on in-situ applications to the vadose zone and the use of dewatering to extend the vadose zone. Bioventing may also be used to introduce methane or other hydrocarbons to stimulate cometabolic degradation of chlorinated compounds, as addressed in Section 5.

3.1.1. Review of the Technology

The first documented evidence of bioventing was reported by the Texas Research Institute, Inc., in a study for the American Petroleum Institute (Texas Research Institute, 1980; 1984). A large-scale model experiment was conducted to test the effectiveness of a surfactant treatment to enhance the recovery of spilled gasoline. Only 30 l of the 250 l originally spilled could be accounted for, and thus questions were raised about the fate of the gasoline. Subsequently, a column study was conducted to determine a diffusion coefficient for soil venting. This column study evolved into a biodegradation study that concluded that as much as 38% of the fuel hydrocarbon was biologically mineralized. Researchers concluded that venting would not only remove gasoline by physical means, but also could enhance microbial activity and promote biodegradation of the gasoline (Texas Research Institute, 1980; 1984).

The first actual field-scale bioventing experiments were conducted by van Eyk for Shell Oil. In 1982 at van Eyk's direction, Delft Geotechnics in The Netherlands initiated a series of experiments to investigate the effectiveness of bioventing for treating hydrocarbon-contaminated soils. These studies are reported in a series of papers (Anonymous, 1986; Staatsuitgeverij, 1986; van Eyk and Vreeken, 1988; van Eyk and Vreeken, 1989a; van Eyk and Vreeken, 1989b).

Wilson and Ward (1986) suggested that using air as a carrier for oxygen could be 1,000 times more efficient than using water, especially in deep, hard-to-flood unsaturated zones. They made the connection between soil venting and biodegradation by observing that "soil venting uses the same principle to remove volatile components of the hydrocarbon." In a general overview of the soil venting process, Bennedsen et al. (1987) concluded that soil venting provides large quantities of oxygen to the unsaturated zone, possibly stimulating aerobic degradation. They suggested that water and nutrients would also be required for significant degradation and encouraged additional investigations.

Biodegradation enhanced by soil venting has been observed at several field sites. Investigators claim that at a soil venting site for remediation of gasoline-contaminated soil, significant biodegradation occurred (measured by a temperature rise) when air was supplied. Investigators pumped pulses of air through a pile of excavated soil and observed a consistent rise in temperature, which they attributed to biodegradation. They claimed that the pile was cleaned up during the summer primarily by biodegradation (Conner, 1988). However, they did not control for natural volatilization from the aboveground pile, and not enough data were published to critically verify their biodegradation claim.

Ely and Heffner (1988) of the Chevron Research Company patented a bioventing process. They did not provide their experimental design and data, but they did present their findings graphically. At a site contaminated by gasoline and diesel oil, they observed a slightly higher removal through biodegradation than through evaporation. At a gasoline-contaminated site, results indicated that about two-thirds of the hydrocarbon removed was due to volatilization and one-third due to biodegradation. At a site containing only fuel oils, approximately 75 l/well/day were biodegraded, while removals by volatilization were low due to low vapor pressures of the fuel oil. Ely and Heffner claimed that the process is more advantageous than strict soil venting because removal is not dependent only on vapor pressure. In the examples stated in the patent, CO_2 was maintained between 6.8% and 11% and O_2 between 2.3% and 11% in vented air. The patent suggested that the addition of water and nutrients may not be acceptable because of flushing of the contaminants to the water table, but nutrient addition is included as part of the patent. The patent recommends flow rates between 50 and 420 m^3/min per well and states that air flows higher than those required for volatilization may be optimum for biodegradation. The Chevron patent is the only patent directly related to the bioventing process. However, other soil venting patents may be relevant.

At Traverse City, Michigan, researchers from the Robert S. Kerr Environmental Research Laboratory (U.S. Environmental Protection Agency) observed a decrease in the toluene concentration in unsaturated zone soil gas, which they measured as an indicator of fuel contamination in the unsaturated zone. They assumed that advection had not occurred, and attributed the toluene loss to biodegradation. The investigators concluded that because toluene concentrations decreased near the oxygenated ground surface, soil venting is an attractive remediation alternative for biodegrading light volatile hydrocarbon spills (Ostendorf and Kampbell, 1989).

This work was followed by a field-scale bioventing pilot study (Kampbell et al., 1992a, Kampbell et al., 1992b; The Traverse Group Inc., 1992). Two experimental configurations were evaluated. In one plot, air was injected near the water table in one row of wells, extracted in another row near the water table, and reinjected at an intermediate depth in a third row of wells. In the second plot, air was injected only at the water table. The plots were thirty feet wide and fifty feet long. The water table was 15 feet below land surface. Air was supplied at 5 cubic feet per minute, resulting in an average residence time of air in the unsaturated zone of 24 hours.

After one year of operation, hydrocarbon vapor concentrations in the unsaturated zone were reduced from 2,000 mg/l to 14 mg/l. The concentration of hydrocarbon vapors in the air that escaped to the atmosphere was less than 0.5 µg/l throughout the entire demonstration. In the plot with direct injection of air, the mass of gasoline above the water table was reduced from 530 g/m^2 plan surface area to 3.3 g/m^2. Below the water table, the mass of gasoline was reduced from 2450 g/m^2 to 1920 g/m^2. In the plot with reinjection of air, the mass above the water after one year of bioventing was 0.3 g/m^2, and the quantity below the water table was 907 g/m^2.

Although most of the gasoline persisted below the water table, benzene was depleted. After one year of bioventing, ground water in contact with the gasoline contained less than 5 µg/l of benzene.

To date, the best documented full-scale bioventing study was initiated in 1988 at Hill Air Force Base (AFB) in Utah (Dupont et al., 1991; Hinchee and Arthur, 1991; Hinchee et al., 1991). This work was followed by a thoroughly documented field pilot-scale study at Tyndall AFB in Florida (Miller, 1990; Miller et al., 1991).

3.1.2. Maturity of the Technology

Bioventing has been applied for remediation of sites since the early- to mid-1980s in the form of soil venting. This process is also known as soil vacuum extraction, vacuum extraction, soil gas extraction, and in-situ volatilization. At most if not all sites where soils are ventilated, oxygen is supplied and biodegradation is stimulated, and in many cases biodegradation is a significant contributor to remediation. In early applications of soil venting, this was not recognized or documented. More recently, soil venting vendors have begun to monitor and document biodegradation, and some are now designing and operating soil venting systems to optimize biodegradation, either in addition to volatilization or to minimize volatilization.

Various forms of undocumented bioventing have been applied to more than 1000 sites worldwide. However, biodegradation has been confirmed by monitoring at no more than 90% of these sites, and optimized at far fewer.

3.1.3. Repositories of Expertise

A number of groups are involved in bioventing research and application. The separation into groups specializing in research vs. application is based upon the author's impressions. This list is not intended to be all inclusive, but only to be representative of organizations known to the author to be significantly involved in bioventing work. The author knows of many other organizations doing bioventing work, and no doubt other organizations unknown to the author are doing bioventing work.

Bioventing Research

Battelle Memorial Institute
Columbus, Ohio
Contact: Robert E. Hinchee
Phone: (614)424-4698
Fax: (614)424-3667

University of Karlsruhe
Karlsruhe, Germany
Contact: Nils-Christian Lund
Phone (49)0721-594016
Fax: (49)0721-551729

U.S. EPA, Robert S. Kerr
Environmental Research Laboratory
Ada, Oklahoma
Contact: John T.Wilson
Phone: (405)436-8532
Fax: (405)436-8529

U.S. EPA, Risk Reduction
Engineering Laboratory
Cincinnati, Ohio
Contact: Richard C. Brenner
Phone: (513)569-7657
Fax: (513)569-7787

U.S. Air Force Civil
Engineering Support Agency
Contact: Catherine M. Vogel
Phone: (904)283-6036
Fax: (904)283-6004

U.S. Navy Civil
Engineering Laboratory
Port Hueneme, CA
Contact: Ronald Hoeppel
Phone: (805)982-1655
Fax: (805)982-1409 or -1418

Utah Water Research Laboratory,
Utah State University
Logan, Utah
Contact: Ryan Dupont
Phone: (801)750-3227
Fax: (801)750-3663

Bioventing Application

Engineering-Science, Inc.
Denver, Colorado
Contact: Doug Downey
Phone: (303)831-8100
Fax: (303)831-8208

Groundwater Technology, Inc.
Trenton, New Jersey
Contact: Richard A. Brown
Phone: (609)587-0300
Fax: (609)587-7908

Integrated Science and Technology
Atlanta, Georgia
Contact: H. James Reisinger, II
Phone: (404)425-3080
Fax: (404)425-0295

The Traverse Group
Ann Arbor, Michigan
Contact: John Armstrong
Phone: (313)747-9300
Fax: (313)483-7532

U.S. Air Force Center for
Environmental Excellence
Brooks AFB, Texas
Contact: Ross N. Miller
Phone: (512)536-4331
Fax: (512)536-4330

Woodward-Clyde Consultants
San Diego, CA
Contact: Desma S. Hogg
Phone: (619)294-9400
Fax: (619)293-7920

TAUW Infra Consult bv
Devemter, The Netherlands
Contact: Leon G.C.M. Urlings
Phone: (31)5700-99528
Fax: (31)5700-99666

Delft Geotechnics
Delft, The Netherlands
Contact: J. van Eyk
Phone: (31)015-693707
Fax: (31)015-610821

3.2. CONTAMINATION THAT IS SUBJECT TO TREATMENT

Bioventing is potentially applicable to any contaminant that is more readily biodegradable aerobically than anaerobically, such as most petroleum hydrocarbons (Atlas, 1981). To date, most applications have been to petroleum hydrocarbons (Hoeppel et al., 1991); however, application to PAHs (Lund et al., 1991; Hinchee and Ong, 1992) and to an acetone, toluene, and naphthalene mixture (Hinchee and Ong, 1992) have been reported.

In most applications, the key is biodegradability vs. volatility. If the rate of volatilization significantly exceeds the rate of biodegradation, removal becomes more of a volatilization process. Figure 3.1 illustrates the general relationship between a compound's physicochemical properties and its potential for bioventing.

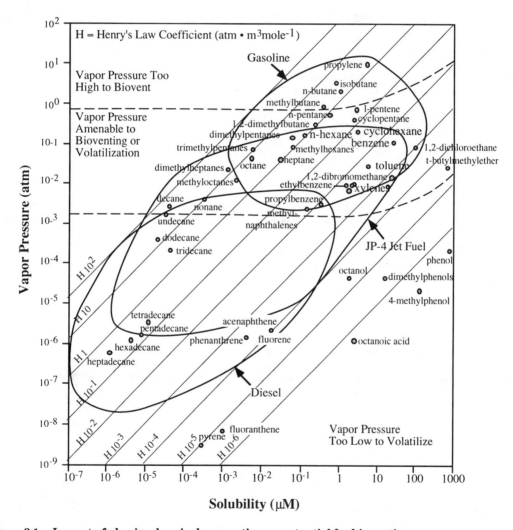

Figure 3.1. Impact of physicochemical properties on potential for bioventing.

In general, compounds with a low vapor pressure (below ~ 1 mm Hg) cannot be successfully removed by volatilization, but can be biodegraded in a bioventing application. Higher vapor pressure compounds, above ~ 760 mm Hg, are gases at ambient temperatures. These compounds volatilize too rapidly to be biodegraded in a bioventing system. Compounds with vapor pressures between 1 and 760 mm Hg may be amenable to either volatilization or biodegradation in a bioventing system. Within this intermediate range lie many of the petroleum hydrocarbon compounds of greatest regulatory interest such as benzene, toluene, and the xylenes. As can be seen in Figure 3.1, various petroleum fuels are more or less amenable to bioventing. Some components of gasoline are too volatile to easily biodegrade. Most of the diesel constituents are sufficiently nonvolatile to preclude volatilization, whereas the constituents of JP-4 jet fuel are intermediate in volatility.

3.3. SPECIAL REQUIREMENTS FOR SITE CHARACTERIZATION

Normal site characterization data are required for implementation of this or any other remedial technology and will not be addressed here. In general, three site characterization tests not typically performed are required for application of bioventing. These tests include: (1) a soil gas survey incorporating measurements of oxygen and carbon dioxide, (2) a pneumatic conductivity test, and (3) an in-situ respiration test. In addition, soil samples should be collected for nutrient analysis, and microbial characterization may be desirable.

3.3.1. Soil Gas Survey

Soil gas surveys are now commonly practiced as part of many site characterizations. The methods for sample collection are well documented in the literature and will not be discussed here (Kerfoot, 1987; Marrin and Kerfoot, 1988). Any method that assures collection of a soil gas sample from discrete depths should be sufficient. Soil gas samples should be analyzed for the contaminant hydrocarbon as well as for oxygen and carbon dioxide. For bioventing to be successful in stimulating biodegradation, the contaminated area must be oxygen deficient. If it is not, the addition of more oxygen will have no effect.

3.3.2. Soil Gas Permeability and Radius of Influence

An estimate of the soil's permeability to air flow (k) and the radius of influence (R_I) of venting wells are both important elements of full-scale bioventing design. On-site testing provides the most accurate estimate of the soil's permeability to air. On-site testing can also be used to determine the radius of influence that can be achieved for a given well configuration, flow rate and air pressure. These data are used to design full-scale systems. Specifically, they are needed to space venting wells, to size blower equipment, and to ensure that the entire site receives a supply of oxygen-rich air to sustain in-situ biodegradation.

The permeability of soils to the flow of gas (k) varies according to grain size, soil uniformity, porosity, and moisture content. The value of k is a physical property of the soil; k does not change with different extraction/injection rates or different pressure levels. Soil gas permeability is generally expressed in the units cm^2 or darcy (1 darcy = 1 x 10^{-8} cm^2). Like hydraulic conductivity, soil gas permeability may vary by more than an order of magnitude on the same site due to soil heterogeneity. The range of typical k values to be expected with different soil types is given in Table 3.1.

The radius of influence (R_I) is defined as the maximum distance from the air extraction or injection well where measurable vacuum or pressure (soil gas movement) occurs. R_I is a function of soil properties, but is also dependent on the configuration of the venting well and extraction or injection flow rates, and is altered by soil stratification. On sites with shallow contamination, the radius of influence can also be increased by impermeable surface barriers such as asphalt or concrete. These paved surfaces may or may not act as vapor barriers. Without a tight seal to the natural soil surface, the pavement will not significantly impact soil gas flow.

TABLE 3.1. DARCY VELOCITY IN RELATION TO SOIL TYPE[a]

Soil Type	k in Darcy
Coarse Sand	100 - 1000
Medium Sand	1 - 100
Fine Sand	0.1 - 1.0
Silts/Clays	<0.1

[a]Source: Johnson et al. (1990)

Several field methods have been developed for determining soil gas permeability (Sellers and Fan, 1991). The most favored field test method is probably the modified field drawdown method developed by Paul Johnson of Shell Development Company (Johnson, 1991). This method involves the injection or extraction of air at a constant rate from a single venting well while measuring the pressure/vacuum changes over time at several monitoring points in the soil away from the venting well.

3.3.3. In-Situ Respiration

An on-site, in-situ respiration test was developed by Hinchee and associates at Battelle (Hinchee and Ong, 1992). The test has been used at numerous sites throughout the United States. To conduct the test, narrowly screened soil gas monitoring points are placed into the unsaturated zone. The soils are vented with air containing an inert tracer gas for a given period of time. The apparatus for the respiration test is illustrated in Figure 3.2.

In a typical experiment, two monitoring point locations -- the test location and a background control location -- are used. A cluster of three to four probes is usually placed in the contaminated soil of the test location. A 1-to-3% concentration of inert gas is added to the air, which is injected for about 24 hours. The air provides oxygen to the soil, while inert gas measurements provide data on the diffusion of O_2 from the ground surface and the surrounding soil and assure that the soil gas sampling system does not leak. A background control location is placed in an uncontaminated site with air injection to monitor natural background respiration.

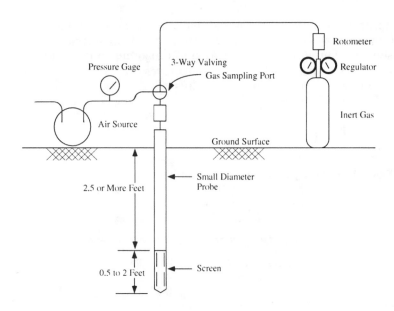

Figure 3.2.   Gas injection/soil gas sampling monitoring point used by Hinchee and Ong (1992) in their in-situ respiration studies.

Measurements of CO_2 and O_2 concentrations in the soil gas are taken before any air and inert gas injection. After air and inert gas injection are turned off, CO_2 and O_2 and inert gas concentrations are monitored over time. The monitoring points in contaminated soil at sites amenable to bioventing show a significant decline in O_2 over a 40- to 80-hour monitoring period. Figure 3.3 illustrates the average results from four such sites, along with the corresponding O_2 utilization rates in terms of percent of O_2 consumed per hour. In general, little or no O_2 utilization was measured in the uncontaminated background monitoring point.

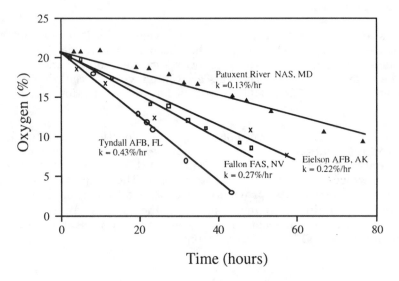

Figure 3.3.   Average oxygen utilization rates measured at four test sites.

The biodegradation rates measured by the in-situ respiration test appear to be representative of those for a full-scale bioventing system. Miller (1990) conducted a 9-month bioventing pilot project at Tyndall AFB at the same time Hinchee and Ong (1992) were conducting their in-situ respiration test. The O_2 utilization rates (Miller, 1990), measured from nearby active treatment areas, were virtually identical to those measured in the in-situ respiration test.

CO_2 production proved to be a less useful measure of biodegradation than O_2 disappearance. The biodegradation rate in milligrams of hexane-equivalent/kilograms of soil per day based on CO_2 appearance is usually less than can be accounted for by the O_2 disappearance. The Tyndall AFB site was an exception. That site had low-alkalinity soils and low-pH quartz sands, and CO_2 production actually resulted in a slightly higher estimate of biodegradation (Miller, 1990). In the case of the higher pH and higher alkalinity soils at Fallon NAS and Eielson AFB, little or no gaseous CO_2 production was measured (Hinchee and Ong, 1992). This could be due to the formation of carbonates from the gaseous evolution of CO_2 produced by biodegradation at these sites. A similar problem was encountered by van Eyk and Vreeken (1988) in their attempt to use CO_2 evolution to quantify biodegradation associated with soil venting. As a rule, O_2 utilization is a more reliable measure of bioventing-induced biodegradation than is CO_2 consumption.

3.4. IMPACT OF SITE CHARACTERISTICS ON APPLICABILITY

Assuming contaminants are present that are amenable to bioventing, gas permeability is probably the most important site characteristic. Soils must be sufficiently permeable to allow movement of enough gas to provide adequate oxygen for biodegradation. Gas permeability is a function of both soil structure and grain size, as well as of soil moisture content. Even in a coarse sand, if soil moisture content is high, adequate gas flow may not be possible. The site must be sufficiently permeable to allow a minimum of approximately one soil gas exchange per week. Typically, permeability in excess of 1 darcy is adequate. When the permeability falls below ~0.1 darcy, gas flow is primarily through either secondary porosity (such as fractures) or any more permeable strata that may be present (such as thin sand lenses).

The feasibility of bioventing in these low-permeability soils is a function of the distribution of flow paths and diffusion of air to and from the flow paths within the contaminated area. In a soil with reasonably good diffusion, a maximum separation of 2 to 4 feet between flow path and contaminant may still result in treatment. This is obviously a very site-specific characteristic. Bioventing has been successful in some low-permeability soils, a silty clay site at Fallon NAS in Nevada (Hinchee, unpublished data), and a silty site on Eilson AFB in Alaska (Leeson et al., 1992). At a clay site on Tinker AFB in Oklahoma, there has been less success (Hinchee and Ong, 1992).

In addition to gas permeability, hydraulic conductivity may be important if it is necessary to either add nutrients or dewater a site.

Another important site characteristic is contaminant distribution. Bioventing is primarily a vadose zone treatment process. The vadose zone may be extended through dewatering, but contamination below the water table cannot be treated.

3.5.   PROCESS PERFORMANCE

In contrast to soil vacuum extraction, which maximizes the flow of air to speed removal of the contaminant, bioventing to enhance biodegradation provides flow of air through hydrocarbon-contaminated soils at rates and configurations that will ensure adequate oxygenation for aerobic biodegradation, and will minimize or eliminate the production of a hydrocarbon-contaminated off-gas.   The addition of nutrients and moisture may be desirable to increase biodegradation rates; however, field research to date indicates that these additions may not always be needed (Dupont et al., 1991; Miller et al., 1991). If necessary, nutrient and moisture addition could take any of a variety of configurations.

The supply of oxygen to contamination in the capillary fringe or below the water table may not be adequate. Dewatering may at times be necessary, depending on the distribution of contaminants relative to the normal water table.

An important feature of good bioventing design is the use of narrowly screened soil gas monitoring points to sample gas in short vertical sections of the soil. These points are utilized to monitor local oxygen concentrations, as oxygen levels in the vent well are not representative of local conditions.

Typically, a soil venting system is installed to draw air from a vent well in the area of greatest contamination. This configuration allows straightforward monitoring of the off-gases. However, its disadvantage is that hydrocarbon off-gas concentrations are maximized and could require permitting and treatment.

Figure 3.4 shows a bioventing system that involves air injection only. Although this is the lowest cost configuration, careful consideration must be given to the fate of injected air. The objective is for most, if not all, of the hydrocarbons to be degraded, and for CO_2 to be emitted at some distance from the injection point. If a building or subsurface structure were to exist within the radius of influence of the well, hydrocarbon vapors could be forced into that structure. Thus, protection of subsurface structures may be required. A bioventing system with this configuration was installed at Hill AFB in 1991 and is currently under study by U.S. EPA RREL and the U.S. Air Force.

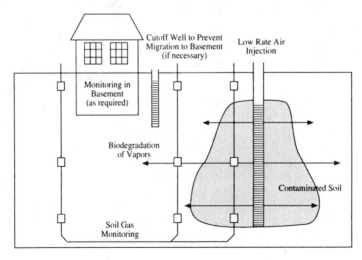

Figure 3.4.   **Conceptual layout of bioventing process with air injection only.**

Figure 3.5 is an illustration of a configuration in which air is injected (the injection may also be by a passive well) into the contaminated zone and withdrawn from clean soils. This configuration allows the more volatile hydrocarbons to degrade prior to being withdrawn, thereby eliminating contaminated off-gases. This configuration typically does not require an air emission permit, although site-specific exceptions may apply. Work with a configuration similar to this at a gasoline site near Atlanta, Georgia, currently is under way.

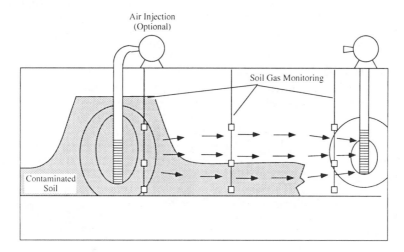

Figure 3.5.   Conceptual layout of bioventing process with air withdrawn from clean soil.

Figure 3.6 illustrates a configuration that may alleviate the threat to subsurface structures while achieving the same basic effect as air injection alone. In this configuration, soil gas is extracted near the structure of concern and reinjected at a safe distance. If necessary, makeup air can be added before injection. Implementation of a configuration similar to this is occurring on a site at Eglin AFB in Florida.

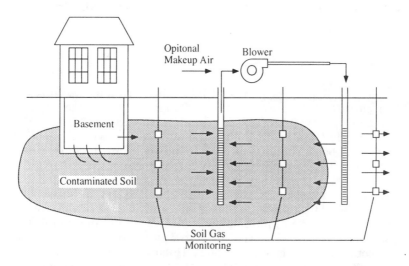

Figure 3.6.   Conceptual layout of bioventing process with soil gas reinjection.

Figure 3.7 illustrates a conventional soil venting configuration at sites where hydrocarbon emissions to the atmosphere are not a problem. This may be the preferred configuration. Dewatering, nutrient, and moisture additions are also illustrated. Dewatering will allow more effective treatment of deeper soils. The optimal configuration for any given site will, of course, depend on site-specific conditions and remedial objectives.

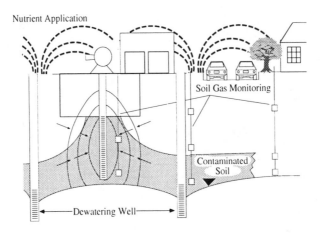

Figure 3.7.   **Conceptual layout of bioventing process with air injection into contaminated soil, coupled with dewatering and nutrient application.**

3.5.1. Case Study: Hill AFB Site

A spill of approximately 100,000 l of JP-4 jet fuel occurred when an automatic overflow device failed at Hill AFB in Ogden, Utah. Contamination was limited to the upper 20 m of a delta outwash of the Weber River. This surficial formation extends from the surface to a depth of approximately 20 m and is composed of mixed sand and gravel with occasional clay stringers. Depth to regional ground water is approximately 200 m; however, water may occasionally be found in discontinuous perched zones. Soil moisture averaged less than 6% in the contaminated soils.

The collected soil samples had JP-4 fuel concentrations up to 20,000 mg/kg, with an average concentration of approximately 400 mg/kg (Oak Ridge National Laboratory, 1989). Contaminants were unevenly distributed to depths of 20 m. Vent wells were drilled to approximately 20 m below the ground surface and were screened from 3 to 18 m below the surface. A background vent was installed in an uncontaminated location in the same geological formation approximately 200 m north of the site.

Venting was initiated in December 1988 by air extraction at a rate of ~40 m³/hr. The off-gas was treated by catalytic incineration, and it was initially necessary to dilute the highly concentrated gas to remain below explosive limits and within the incinerator's hydrocarbon operating limits. The venting rate was gradually increased to ~2,500 m³/hr as hydrocarbon concentration levels dropped. During the period between December 1988 and November 1990, more than 1.0 x 10⁶ m³ of soil gas was extracted from the site. In

November 1989, ventilation rates were reduced to between ~500 and 1000 m³/hr to provide aeration for bioremediation while reducing off-gas generation. This change allowed removal of the catalytic incinerator, saving ~$6,000 per month.

During extraction, oxygen and hydrocarbon concentrations in the off-gas were measured. To quantify the extent of biodegradation at the site, the oxygen was converted to an equivalent basis. This was based on the stoichiometric oxygen requirement for hexane mineralization. JP-4 hydrocarbon concentrations were determined based on direct readings of a total hydrocarbon analyzer calibrated to hexane. Based on these calculations, the mass of the JP-4 fuel as carbon removed was ~50,000 kg volatilized and 40,000 kg biodegraded. Figures 3.8 and 3.9 illustrate these results.

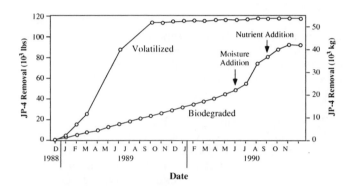

Figure 3.8. Cumulative hydrocarbon removal from the Hill AFB Building 914 soil venting site (Dupont, et al., 1991).

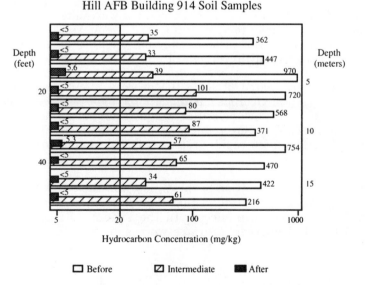

Figure 3.9. Results of soil analysis at Hill AFB before and after venting. (Each bar represents the average of 14 or more samples) (Dupont, et al., 1991).

Hinchee and Arthur (1991) conducted bench-scale studies using soils from this site and found that, in the laboratory, both moisture and nutrients became limiting after aerobic conditions were achieved. This led to the addition of first moisture and then nutrients in the field. The results of these field additions are shown in Figure 3.8. Moisture addition clearly stimulated biodegradation; nutrient addition did not.

The failure to observe an effect of nutrient addition could be explained by a number of factors, including:

- The nutrients failed to move in the soils; this is a problem particularly for ammonia and phosphorus (Aggarwal et al., 1991).
- Remediation of the site was entering its final phase, and there was not enough time for the microbes to respond to the nutrient addition.
- Nutrients may not have been limiting.

3.5.2. Case Study: Tyndall AFB Site

As a follow-up to the Hill AFB research, a more controlled study was designed at Tyndall AFB (Miller et al., 1991). The experimental area in this study was located at a site where past JP-4 fuel storage had resulted in contaminated soils. The nature and volume of fuel spilled or leaked were unknown. The site soils are a fine- to medium-grained quartz sand. The depth to ground water was 0.5 to 1.5 m.

Four test cells were constructed to allow control of gas flow, water flow, and nutrient addition. Test cells V1 and V2 were installed in the hydrocarbon-contaminated zone; the other two were installed in uncontaminated soils. Initial site characterization indicated the mean soil hydrocarbon levels were 5,100 and 7,700 mg of hexane-equivalent/kg in treatment plots V1 and V2, respectively. The contaminated area was dewatered, and hydraulic control was maintained to keep the depth to water at ~2 m. This exposed more of the contaminated soil to aeration. During normal operation, airflow rates were maintained at approximately one air-filled void volume per day.

Biodegradation and volatilization rates were much higher at the Tyndall AFB site than those observed at Hill AFB; these higher rates were likely due to higher average levels of contamination, warmer temperatures, and the presence of moisture. After 200 days of aeration, an average hydrocarbon reduction of ~2,900 mg/kg was observed. This represents a reduction in total hydrocarbons of approximately 40%.

The study was terminated because the process monitoring objectives had been met; biodegradation was still vigorous. Although the total petroleum hydrocarbons had been reduced by only 40%, the low-molecular-weight aromatics -- benzene, toluene, ethylbenzene, and xylenes (BTEX) -- were reduced by more than 90% (Figure 3.10). It appears that the bioventing process more rapidly removes BTEX compounds than other JP-4 fuel constituents.

Another important observation of this study was the effect of temperature on the biodegradation rate. Miller (1990) found that the van Hoff-Arrhenius equation provided an excellent model of temperature effects. In the Tyndall AFB study, soil temperature varied by only ~7°C, yet biodegradation rates were approximately twice as high at 25°C than at 18°C.

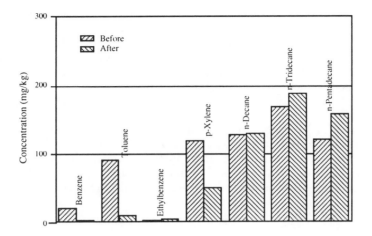

Figure 3.10. **Results of soil analysis from Plot V2 at Tyndall AFB before and after venting, (Each bar represents the average of 21 or more soil samples. Miller et al., 1991).**

In the Tyndall AFB study, the effects of moisture and nutrients were observed in a field test. Two side-by-side plots received identical treatment, except that one (V2) received both moisture and nutrients from the initiation of the study while the other plot (V1) received neither for 8 weeks, then moisture only for 14 weeks, followed by both moisture and nutrients for 7 weeks. As illustrated in Figure 3.11, no significant effect of moisture or nutrients was observed. The lack of moisture effect contrasts with the Hill AFB findings, but is most likely the result of contrasting climatic and hydrogeologic conditions. Hill AFB is located on a high-elevation desert with a very deep water table. Tyndall AFB is located in a moist subtropical environment, and at the site studied, the water table was maintained at a depth of approximately 2 m.

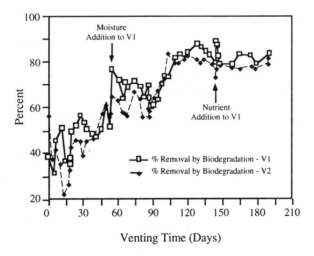

Figure 3.11. **Cumulative percent hydrocarbon removal at Tyndall AFB for Sites V1 and V2 (Miller et al., 1991).**

The nutrient findings support field observations at Hill AFB that the addition of nutrients does not stimulate biodegradation. Based on acetylene reduction studies, Miller (1990) speculates that adequate nitrogen was present due to nitrogen fixation. Both the Hill and Tyndall AFB sites were contaminated for several years before the bioventing studies, and both sites were anaerobic. It is possible that nitrogen fixation, which is maximized under these conditions, provided the required nutrients. In any case, these findings show that nutrient addition is not always required.

In the Tyndall AFB study, a careful evaluation of the relationship between air flow rates and biodegradation and volatilization was made. It was found that extracting air at the optimal rate for biodegradation resulted in 90% removal by biodegradation and 10% removal by volatilization. It was also found that the volatilized contaminants were completely biodegraded after the air was passed through clean soil.

3.5.3. Performance of Other Sites

In addition to the Hill AFB and Tyndall AFB sites, numerous other site studies are reported in the literature. A summery of several other studies is presented in Table 3.2.

TABLE 3.2.   COMPARISON OF BIODEGRADATION RATES OBTAINED BY THE IN-SITU RESPIRATION TEST WITH OTHER STUDIES

Site	Scale of Application	Respiration Contaminants	Estimated Rates (% O_2/hour)	Biodegradation Rates	References
Various (8 locations)	In situ respiration tests	Various	0.02 - 0.99	0.4 - 19 mg/kg/day	Hinchee and Ong, 1992
Hill AFB, Utah	Full-scale, 2 years	JP-4 Jet Fuel	up to 0.52	up to 10 mg/kg/day[a]	Hinchee et al., 1991
Tyndall AFB, Florida	Field pilot, 1 year and in situ respiration tests	JP-4 Jet Fuel	0.1 - 1.0	2 - 20 mg/kg/day	Miller, 1990
Netherlands	Undefined	Undefined	0.1 - 0.26	2- 5 mg/kg/day[b]	Urlings et al., 1990
Netherlands	Field pilot, 1 year	Diesel	0.42	8 mg/kg/day	van Eyk and Vreeken, 1989b
Undefined	Full-scale	Gasoline and Diesel	--	50 kg/well/day[c]	Ely and Heffner, 1988
Undefined	Full-scale	Diesel	--	100 kg/well/day[c]	Ely and Heffner, 1988
Undefined	Full-scale	Fuel Oil	--	60 kg/well/day[c]	Ely and Heffner, 1988
New Zealand	Pilot-scale/ Full-scale	Diesel Spent Oil	--	0.2 - 20 mg/kg/day	Hogg et al., 1992
Traverse City	Pilot-Scale	Aviation Gasoline	--	4 mg/kg/day[b]	Wilson, 1992

a   Rates reported by Hinchee et al. (1991) were first order with respect to oxygen; for comparison purposes, these have been converted to zero order with respect to hydrocarbons at an assumed oxygen concentration of 10 percent.
b   Rates reported as oxygen consumption rates; these have been converted to hydrocarbon degradation rates assuming a 3:1 oxygen-to-hydrocarbon ratio.
c   Units are in kilograms of hydrocarbon degraded per 30 standard cubic feet per minute (scfm) extraction vent well per day.

3.6. PROBLEMS ENCOUNTERED WITH THE TECHNOLOGY

The primary problems encountered with bioventing are:

- Accurate estimate of emissions -- One of the key variables in bioventing cost and design is the need or lack of need for off-gas treatment. To permit an emission with or without off-gas treatment, regulators typically require an estimate of emission rate.

- Time required for remediation -- Bioventing that is primarily dependent on biodegradation is a slow process requiring two or more years for remediation. At many sites this can be a problem.

- Determination of effectiveness on nonpetroleum hydrocarbons -- Many sites are contaminated with a mixture of chemical wastes, and little is known of the effectiveness of bioventing on nonpetroleum hydrocarbons.

- Regulatory acceptance -- With this technology, as with many emerging technologies, obtaining regulatory acceptance can be difficult.

3.7. COSTS

As with many emerging technologies, not much published experience exists to precisely determine cost; however, some general guidelines are available. The basic cost of soil venting equipment has been estimated to be as low as $10 per yd^3. Off-gas treatment costs can run $30 to 50/yd^3 (Long, 1992). The key to the cost of bioventing is the monitoring costs, and these are very site specific. Due to the relatively long time period required for biodegradation, a bioventing site optimized for biodegradation as opposed to volatilization will incur higher monitoring costs. On some sites this may be offset by not having gas treatment costs, thus reducing total remediation costs.

3.8. REGULATORY ACCEPTANCE

The primary obstacle to regulatory acceptance is demonstrating potential effectiveness. Many regulators are unfamiliar with the technology and skeptical of accepting a remedial approach that may require 2 to 3 years to show results. Generally after the responsible regulator(s) agree to allow the use of bioventing, permitting is not difficult.

3.9. KNOWLEDGE GAPS AND RESEARCH OPPORTUNITIES

Bioventing has been performed and monitored at several field sites contaminated with middle distillate fuels, mainly JP-4 jet fuel. Yet the effects of environmental variables on bioventing treatment rates are not well understood. In-situ respirometry at additional sites with drastically different geologic conditions has further defined environmental limitations and site-specific factors that are pertinent to successful bioventing. However, the relationship between respirometric data and actual bioventing treatment rates have not been clearly determined. Additional field respirometry and closely monitored field pilot bioventing studies at the same sites are needed to determine what types of contaminants can be successfully treated in situ by bioventing and what the environmental limitations are.   Studies are also needed on a wide variety of contaminants.   Studies to date clearly show that many preconceptions regarding the factors that control bioventing rates may be incorrect.   For example, active respiration at a subarctic site at Eielson AFB near Fairbanks, Alaska, suggests that good hydrocarbon degradation can occur in situ at locations that are continually subjected to a cold environment.   Failure to accelerate biodegradation rates by adding nitrogen fertilizer to biovented soils that contain low nitrogen levels indicates that nutrient addition at some sites may not be required.   Also, fine-grained moist clayey soils have been readily aerated and showed aerobic respiration, indicating that bioventing may be feasible in soils having low permeabilities.   Other low permeability sites have not proven amenable to bioventing, and better procedures to evaluate sites are needed.

Vapor phase biodegradation occurs and can take place in situ.   The question of how soil sorption and partitioning of volatile organic compounds into soil air affects biodegradation rates was addressed earlier by McCarty (1987).   This question needs further attention as the movement of the vapor phase in soils is complex and dependent on changing soil environmental conditions.

Bioventing rates need to be determined under varying vapor extraction rates since an important purpose for bioventing is to biodegrade the vapor within the soil profile.   Minimal soil aeration levels that provide for high degradation rates must be determined under different soil conditions.   Interaction of the vapor phase with soil particles and microorganisms in the uncontaminated soil profile needs further research in both the laboratory and in the field.

REFERENCES

Aggarwal, P.K., J.L. Means, and R.E. Hinchee. 1991. Formulation of nutrient solutions for in situ bioremediation. In: *In Situ Bioreclamation. Applications and Investigations for Hydrocarbon and Contaminated Site Remediation.* Eds., R.E. Hinchee and R.F. Olfenbuttel. Butterworth-Heinemann. Stoneham, Massachusetts. pp. 51-66.

Anonymous. 1986. In situ reclamation of petroleum contaminated sub-soil by subsurface venting and enhanced biodegradation. *Research Disclosure.* No. 26233, 92-93.

Atlas, R.M. 1981. Microbial degradation of petroleum hydrocarbons: An environmental perspective. *Microbiol. Rev.* 45: 180-209.

Bennedsen, M.B., J.P. Scott, and J.D. Hartley. 1987. Use of vapor extraction systems for in situ removal of volatile organic compounds from soil. In: *Proceedings of National Conference on Hazardous Wastes and Hazardous Materials.* Washington, DC. pp. 92-95.

Conner, J.S. 1988. Case study of soil venting. *Poll. Eng.* 7: 74-78.

Dupont, R.R., W. Doucette, and R.E. Hinchee. 1991. Assessment of in situ bioremediation potential and the application of bioventing at a fuel-contaminated site. In: *In Situ and On-Site Bioreclamation.* Eds., R.E. Hinchee and R.F. Olfenbuttel. Butterworth-Heinemann. Stoneham, Massachusetts. pp. 262-282.

Ely, D.L., and D.A. Heffner. 1988. Process for in-situ biodegradation of hydrocarbon contaminated soil. U.S. Patent Number 4,765,902.

Hinchee, R.E., and M. Arthur. 1991. Bench-scale studies of the soil aeration process for bioremediation of petroleum hydrocarbons. *J. Appl. Biochem. Biotech.* 28/29: 901-906.

Hinchee, R.E., D.C. Downey, R.R. Dupont, P. Aggarwal, and R.N. Miller. 1991. Enhancing biodegradation of petroleum hydrocarbon through soil venting. *J. Hazardous Materials.* 27: 315-325.

Hinchee, R.E., and S.K. Ong. 1992. A rapid in situ respiration test for measuring aerobic biodegradation rates of hydrocarbons in soil. Submitted to the *Journal of the Air & Waste Management Association.* 21 pp.

Hoeppel, R. E., R.E. Hinchee, and M.R. Arthur. 1991. Bioventing soils contaminated with petroleum hydrocarbons. *J. Industrial Microbiology.* 8:141-146.

Hogg, D.S., R.J. Burden, and P.J. Riddell. 1992. In situ vadose zone bioremediation of soil contaminated with non-volatile hydrocarbon. Presented at HMCRI Conference. February 4. San Francisco, California.

Johnson, P.C., M.W. Kemblowski, and J.D. Colthart. 1990. Quantitative analysis for the cleanup of hydrocarbon-contaminated soils by in-situ soil venting. *Ground Water.* 28(3):413-429. May-June.

Johnson, P.C. 1991. *HyperVentilate Users Manual.* Shell Development. Houston, Texas. 3 pp.

Kampbell, D.H., J.T. Wilson, and C.J. Griffin. 1992a. Bioventing of a gasoline spill at Traverse City, Michigan. In: *Bioremediation Of Hazardous Wastes.* EPA/600/R-92/126. Office of Research and Development.

Kampbell, D.H., J.T. Wilson, C.J. Griffin, and D.W. Ostendorf. 1992b. Bioventing reclamation pilot project--aviation gasoline spill. In: *Abstracts, Subsurface Restoration Conference.* National Center for Ground Water Research. June 21-24, 1992. Dallas, Texas. p. 297.

Kerfoot, H.B. 1987. Soil-gas measurement for detection of groundwater contamination by volatile organic compounds. Environ. Sci. Technol. 21(1):1022-1024.

Leeson, A., R.E. Hinchee, J. Kittle, G. Sayles, C.M. Vogel, and R.N. Miller. 1992. Optimizing bioventing in shallow vadose zones and cold climates. Submitted for Publication in the *Proceedings of the In-Situ and On-Site Bioremediation Symposium.* Niagra on the Lake, Canada.

Long, G. 1992. Bioventing and vapor extraction: Innovative technologies for contaminated site remediation. *J. Air and Waste Mgt. Assoc.* 43(3):345-348.

Lund, N.-Ch., J. Swinianski, G. Gadehus, and D. Maier. 1991. Laboratory and field tests for a biological in situ remediation of a coke oven plant. In: *In Situ Bioreclamation. Applications and Investigations for Hydrocarbon and Contaminated Site Remediation.* Eds., R.E. Hinchee and R.F. Olfenbuttel. Butterworth-Heinemann. Stoneham, Massachusetts. pp. 396-412.

Marrin, D.L. and W.B. Kerfoot. 1988. Soil gas surveying techniques. Environ. Sci. Technol. 22(7):740-745.

McCarty, P.L. 1987. Bioengineering issues related to in situ remediation of contaminated soils and groundwater. In: *Proceedings, Conference on Reducing Risk from Environmental Chemicals Through Biotechnology.* Seattle, Washington. July.

Miller, R.N. 1990. A field scale investigation of enhanced petroleum hydrocarbon biodegradation in the vadose zone combining soil venting as an oxygen source with moisture and nutrient additions. Ph.D. Dissertation. Utah State University. Logan, Utah.

Miller, R.N., R.E. Hinchee, and C. Vogel. 1991. A field-scale investigation of petroleum hydrocarbon biodegradation in the vadose zone enhanced by soil venting at Tyndall AFB, Florida. In: *In Situ Bioreclamation. Applications and Investigations for Hydrocarbon and Contaminated Site Remediation.* Eds., R.E. Hinchee and R. F. Olfenbuttel. Butterworth Publishers. Stoneham, Massachusetts. pp. 283-302.

Oak Ridge National Laboratory. 1989. Soil characteristics: Data summary, Hill Air Force Base Building 914 fuel spill soil venting project. An unpublished report to the U.S. Air Force.

Ostendorf, D.W., and D.H. Kampbell. 1989. Vertical profiles and near surface traps for field measurement of volatile pollution in the subsurface environment. In: *Proceedings of NWWA Conference on New Techniques for Quantifying the Physical and Chemical Properties of Heterogeneous Aquifers*. National Water Well Association. Dublin, Ohio.

Sellers, K., and C.Y. Fan. 1991. Soil vapor extraction: Air permeability testing and estimation methods. In: *Proceedings of the 17th RREL Hazardous Waste Research Symposium*. EPA/600/9-91/002, April.

Staatsuitgeverij. 1986. Proceedings of a Workshop, 20-21 March, 1986. *Bodembeschermingsreeeks* No. 9. *Biotechnologische Bodemsanering*. pp. 31-33. Rapportnr. 851105002. ISBN 90-12-054133. Ordernr. 250-154-59. Staatsuitgeverij Den Haag: The Netherlands.

Texas Research Institute. 1980. *Laboratory Scale Gasoline Spill and Venting Experiment*. American Petroleum Institute. Interim Report No. 7743-5:JST.

Texas Research Institute. 1984. *Forced Venting to Remove Gasoline Vapor from a Large-Scale Model Aquifer*. American Petroleum Institute. Final Report No. 82101-F:TAV.

The Traverse Group, Inc. 1992. *Bioventing Reclamation Pilot Program. U.S. Coast Guard Air Station. Traverse City, Michigan. Final Report*. Prepared for the U.S. Coast Guard and the U.S. EPA (Robert S. Kerr Environmental Research Laboratory).

van Eyk, J., and C. Vreeken. 1988. Venting-mediated removal of petrol from subsurface soil strata as a result of stimulated evaporation and enhanced biodegradation. *Med. Fac. Landbouww. Riiksuniv. Gent.* 53(4b): 1873-1884.

van Eyk, J., and C. Vreeken. 1989a. Model of petroleum mineralization response to soil aeration to aid in site-specific, in situ biological remediation. In: *Groundwater Contamination: Use of Models in Decision-Making, Proceedings of an International Conference on Groundwater Contamination*. Ed., G. Jousma. Kluwer Boston/London. pp. 365-371.

van Eyk, J., and C. Vreeken. 1989b. Venting-mediated removal of diesel oil from subsurface soil strata as a result of stimulated evaporation and enhanced biodegradation. In: *Hazardous Waste and Contaminated Sites, Envirotech Vienna*. Vol. 2, Session 3. ISBN 389432-009-5. Westarp Wiss., Essen. pp. 475-485.

Wilson, J.T. 1992. Technologies for contaminant destruction: Enhanced biological electron acceptor H_2O_2. In: *Abstracts, Subsurface Restoration Conference*. National Center for Ground Water Research. June 21-24, 1992. Dallas, Texas. pp. 86-88.

Wilson, J.T., and C.H. Ward. 1986. Opportunities for bioremediation of aquifers contaminated with petroleum hydrocarbons. *J. Ind. Microbiol.* 27:109-116.

SECTION 4

TREATMENT OF PETROLEUM HYDROCARBONS IN GROUND WATER BY AIR SPARGING

Richard Brown
Groundwater Technology, Inc.
310 Horizon Center Drive
Trenton, New Jersey 08691
Telephone: (609)587-0300
Fax: (609)587-7908

4.1.   INTRODUCTION

Petroleum hydrocarbon contamination of ground water is a significant and often complex problem. The complexity results from the fact that ground-water contamination is a product of soil contamination. Especially with older spills, significant amounts of hydrocarbons can be trapped below the water table. Traditional pump and treat is ineffective because of the low solubility of trapped oily phase hydrocarbons. Typically, less than five percent of a hydrocarbon spill ever enters the dissolved phase. The bulk of contamination is sorbed to soil and/or aquifer solids. Venting is ineffective because of the water saturation. Bioremediation, while effective in treating this trapped hydrocarbon, is often quite expensive when relying on chemical oxygen carriers such as hydrogen peroxide. There has been a need, therefore, to develop a technology which could more cost effectively address petroleum hydrocarbon contamination of ground water. A technology that offers the most promise is air sparging.

Air sparging, simply viewed, is the injection of air under pressure below the water table. This creates a transient air filled porosity by displacing water in the soil matrix (Figure 4.1). The minimum pressure that is required to displace water in an air sparging system is that which is needed to overcome the resistance of the soil matrix to

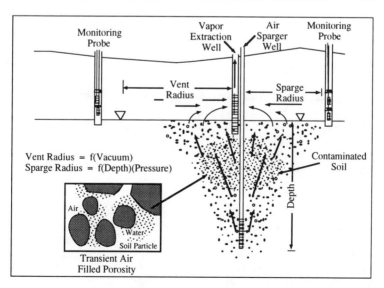

Figure 4.1.   Diagram of air sparging system.

air flow. This resistance to flow is a function of the height of the water column that needs to be displaced and of the flow restriction (air/water permeability) of the soil matrix. When this "break-out" pressure is achieved, air enters the soil matrix, travels horizontally and vertically through the soil matrix displacing water, and eventually exits into the vadose zone.

Air sparging is a relatively new treatment technology for addressing contamination below the water table. By displacing water in the soil matrix and creating a transient air filled porosity, air sparging provides two benefits. First, air sparging enhances biodegradation by increasing oxygen transfer to the ground water. Second, it can enhance the physical removal of organics by direct volatile (vapor phase) extraction.

4.2. DEVELOPMENT OF AIR SPARGING

Soil vapor extraction (SVE, or venting), the inducement of air flow by the application of a vacuum, has long been recognized as one of the more effective means of treating volatile organic compounds (VOCs), particularly petroleum fuels in the vadose zone (Hoag and Marley, 1984). SVE addresses VOCs through two primary mechanisms. First, SVE stimulates biodegradation by supplying oxygen for aerobic metabolic processes. Second, it physically removes the contaminants by removing vapors associated with adsorbed contaminants. Both mechanisms are dependent on the effective movement of air through the subsurface.

Because SVE technology depends on the flow of air through a soil matrix, it has been obviously limited to the treatment of unsaturated soils. SVE cannot **directly** treat VOC-contaminated soils below the water table or contaminated ground water because there is no air filled porosity below the water table, and, therefore, no air to move. SVE has, however, been used to indirectly stimulate the biodegradation of dissolved contaminants by increasing oxygen content in the vadose zone, and therefore diffusion from the vadose zone to ground water (Clayton et al., 1989). To directly treat saturated zone contaminants most effectively with SVE, however, generally requires that the site be effectively dewatered so that a vacuum can be applied and air flow induced.

The difficulty of dewatering and the costs and problems of treating the extracted ground water has made coupled dewatering-SVE systems less than an optimal solution. As a result, there has been a search for technology that could effectively extend the utility of SVE to saturated systems.

Air sparging effectively removes contamination below the water table. Air sparging is the injection of air <u>directly</u> into a saturated formation. Air sparging can successfully treat VOCs and petroleum hydrocarbons in ground-water aquifers through direct volatile removal and biodegradation. Air sparging has been extensively used in Germany since 1985 (Hiller and Gudemann, 1988) and was successfully introduced in the United States in 1990 (Brown et al., 1991; Marley et al., 1990; Middleton and Hiller, 1990).

Air sparging, as practiced today, should not be confused with older systems, which were also called air sparging and were used in early bioremediation projects (Raymond et al., 1975). The difference between the two technologies is where the air is injected. With older technologies the air was injected into the water column in the well. The air, in this case, travels through the water column and does not directly contact the formation matrix. With modern air sparging, the air injection pressure is greater than the hydraulic head, thus the well contains no water and air is directly injected into the formation. Figure 4.2 shows the differences between the two "air sparging" technologies.

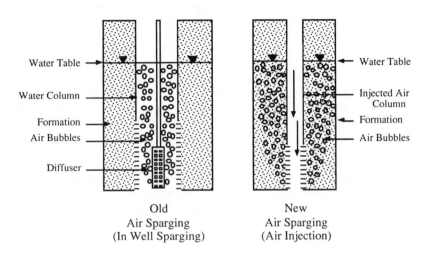

Figure 4.2. **Differences between old and new air sparging technologies.**

Air sparging is an emerging technology for the treatment of ground water contaminated with volatile organic compounds. It is being used to increasingly greater extents to treat petroleum hydrocarbon contaminated ground-water aquifers, overcoming the limitation of SVE for treating saturated zone contaminants and improving the efficacy of bioremediation. The benefits and limitations of this technology are still being defined both in field application and research.

4.3. PRINCIPLES OF THE TECHNOLOGY

When air is injected into a contaminated soil/water matrix (i.e., an aquifer), there are a number of phenomena that result from the air movement. Some are beneficial - they remove contamination; and some are actually or potentially detrimental -they increase or spread contamination. The following is a list and description of these phenomena:

Enhanced Oxygenation: Air traveling through the aquifer dissolves in the soil water and replenishes oxygen that may have been depleted by chemical or biological processes. Normal oxygen replenishment is slow, as it relies on diffusion from the surface of the water table. Sparged air, which is distributed throughout the aquifer, has a short diffusion path length. Enhanced oxygenation is a beneficial phenomenon as it can stimulate biodegradation.

Enhanced Dissolution: Air traveling through the aquifer causes turbulence in the soil pores. This mixes the water and adsorbed VOCs and enhances their partitioning into the water phase. Normal water/soil contact is static and dissolution is diffusion-limited. Enhanced dissolution is beneficial if the ground water is collected, but detrimental if the contaminated plume is not captured or treated (by in-situ stripping). Dissolution can also help promote biodegradation.

Volatilization: Adsorbed phase contaminants will evaporate into the air stream and be carried into the vadose zone. The extent of volatilization is governed by the vapor pressure of the VOC. Volatilization is prevented in normal saturated environments because there is no air phase. Volatilization is a beneficial process as it can remove a significant mass of contaminants.

Ground-water Stripping: The aerated aquifer can act as a crude air stripper if <u>sufficient</u> air flow is passed through the soil matrix. VOCs with a sufficiently high Henry's Law constant will volatilize from the water into the air stream and be removed. This is a generally beneficial process.

Physical Displacement: At very high air flow rates, water can be rapidly and physically displaced. This is observed often in air-rotary drilling. The displaced water, if contaminated, will spread contamination in any direction and is thus not easily captured by existing ground-water systems. Displacement is a generally detrimental phenomenon and should be avoided.

These phenomena are all the result of air passing through the aquifer matrix. Which process is active is generally a function of the amount of air passing through the soil matrix. These phenomena do not all occur at the same air flow rates. As shown in Figure 4.3, oxygenation and dissolution occur at essentially all air flow rates. Volatilization and stripping require moderate rates of air flow. Physical displacement generally only occurs at high pressures or flows. To maximize the benefits of air sparging and minimize the detriments requires an optimization of air flow. Too low an air flow will not effectively remove VOCs and may increase ground-water concentrations; too high a flow can rapidly and physically spread the contamination. Optimizing the air flow will maximize mass removal while minimizing the potential spread of contamination.

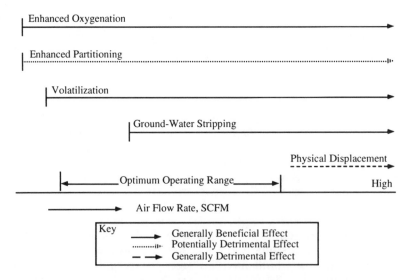

Figure 4.3.   The effects of air flow in saturated environment as a function of air flow rate.

4.4. BENEFITS OF AIR SPARGING

Air sparging is a potentially effective means of treating petroleum hydrocarbons. That is because air sparging promotes two significant removal mechanisms - biodegradation and volatilization.

A primary benefit in treating petroleum hydrocarbons with air sparging is that it is an effective means of supplying oxygen to the saturated zone. This benefit leads to a key application of air sparging, i.e., enhancing aerobic bioremediation. Air sparging results in efficient aeration as a result of several factors. First, there is penetration of air into the contaminated saturated zone. Under normal conditions air only contacts the surface of the aquifer. With air sparging the contact is distributed over the entire sparged interval. Second, because air sparging creates air-filled porosity in the soil matrix, the diffusive path length of air (oxygen) into the water is considerably shortened compared to normal ground-water conditions. Under normal conditions the distance between the air and water phases can be on the order of meters; with air sparging, the distance will be, at most, only several times greater than a soil pore, i.e., a few millimeters. Third, the "turbulence" caused by air sparging enhances the dissolution and distribution of oxygen into the water phase. Since biodegradation is critically dependent on oxygen supply, the efficient aeration engendered by air sparging will enhance bioremediation.

Given this aeration efficiency, an advantage of air sparging is the amount of oxygen that it can provide for biodegradation compared with the use of hydrogen peroxide. Even assuming limited utilization of oxygen in the sparged air, air sparging can supply significant amounts of oxygen for bioremediation. Table 4.1 compares the amount of oxygen supplied to bioremediation from air sparging or from the use of hydrogen peroxide.

TABLE 4.1. OXYGEN AVAILABILITY, lb/day

AIR SPARGING				HYDROGEN PEROXIDE (1000 PPM)			
Flow		*Utilization*		*Flow*		*Utilization*	
SCFM	100%	50%	10%	GPM	100%	50%	10%
10	236	118	24	10	56	28	6
25	590	295	59	25	140	70	14
50	1182	590	118	50	280	140	28

As can be seen, a total sparge flow rate of 25 CFM at only a 10% utilization provides as much oxygen as injecting 10 gpm of 1000 ppm H_2O_2 assuming 100% utilization. One of the problems with hydrogen peroxide, in addition to the relatively low oxygen content at proper use rates, is that peroxide can be quite unstable in some soils. In such cases its utilization is much less than 100% due to premature decomposition. In these situations air sparging would have an even greater advantage as an oxygenation source.

In addition to effective oxygenation, air sparging can also remove contaminants through volatilization, either directly, by "evaporating" the adsorbed phase, or, indirectly, by stripping contaminated ground water.

In the first volatilization process, direct extraction, air bubbles that form during air sparging traverse horizontally and vertically through the soil column, creating transient air-filled regimes in the saturated soil matrix. Volatile hydrocarbons that are exposed to this sparged air environment "evaporate" into the gas phase and are carried by the air stream into the vadose zone where they can be captured by a vent system. Whether a compound is extractable by an air sparge system is, as with soil vapor extraction (SVE), determined by its vapor pressure. The practical vapor pressure limit for an air sparging system, as it is for SVE, is ~1 mm Hg.

In the second volatilization process, ground-water stripping, the key to successful treatment is attaining good contact between the injected air and contaminated ground water. Given good air-to-water contact, the effectiveness of air sparging in ground-water treatment is then determined by the Henry's Law constant of the dissolved VOC contaminants being treated by the air sparger system. A Henry's Law constant, K_H (atm-m^3-mole^{-1}), greater than 10^{-5} indicates a volatile constituent that can be removed by air sparging. Table 4.2 lists the Henry's Law constant for several volatile hydrocarbon constituents.

TABLE 4.2.   HENRY'S CONSTANT FOR SELECTED HYDROCARBONS

CONSTITUENT	HENRY'S CONSTANT, K_H $(atm\text{-}m^3\text{-}mole\text{-}1)$
Cyclohexane	1.9×10^2
Benzene	5.6×10^{-3}
Ethylbenzene	8.7×10^{-3}
Toluene	6.3×10^{-3}
Xylene	5.7×10^{-3}
Naphthalene	4.1×10^{-4}
Phenanthrene	2.5×10^{-5}

The olefins and BTEX compounds are easily stripped from water by air sparging, as indicated by their Henry's Law constants being much greater than 10^{-5} (atm-m^3-mole^{-1}); heavier compounds such as PAHs are more difficult to remove.

When air sparging is applied, the result is a complex partitioning of the petroleum hydrocarbon between the adsorbed, dissolved and vapor state, as well as a complex series of removal mechanisms that may be engendered - removal as a vapor, biodegradation, and removal as a solute in ground water. Which mechanisms are the primary removal mechanisms and which are secondary depends on the volatility of the contaminant. As shown in Figure 4.4, with a highly volatile product, the primary partitioning is into the vapor state, and the primary removal mechanism is through volatilization. By contrast, with a low volatility product, partitioning is primarily to the adsorbed or dissolved state, and the primary removal mechanism is through biodegradation.

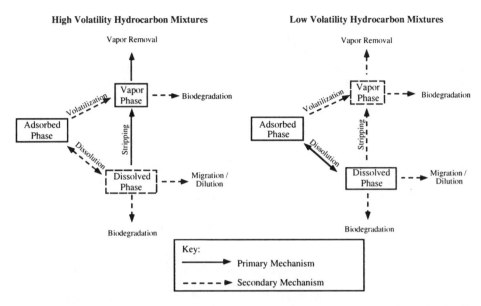

Figure 4.4.   **Air sparging partitioning and removal mechanisms as a function of volatility.**

Because air sparging can both stimulate biodegradation as well as remove hydrocarbon vapors, it can treat a wide range of petroleum hydrocarbon products. As shown in Figure 4.5, the treatment mechanisms, however, vary with the type of product. Heavy products such as No. 6 fuel oil are treated primarily through biodegradation. Light products such as gasoline are treated more through simple volatilization than through biodegradation.

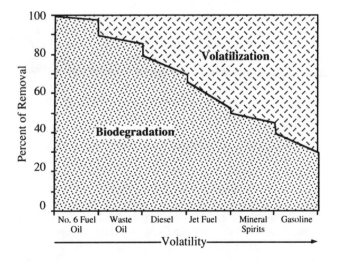

Figure 4.5.   **Air sparging removal mechanisms as a function of product volatility.**

4.5.   DANGERS OF AIR SPARGING

A fundamental issue in any remedial process is having control over the process. With processes that are based on extraction such as SVE or ground-water recovery, the process begins with the system under control because contaminants are being drawn to a point of collection.   By contrast, injection systems such as air sparging start with no control because flow is away from the injection point.   Control must be **gained** and **maintained**. Therefore, with air sparging, anything that affects the control of the flow of air can limit the application of air sparging.   There are two fundamental concerns with the efficacy and utilization of air sparging. These concerns may be categorized as structural and operational.

First, with respect to structural concerns, air sparging is based on the controlled injection of air into a saturated soil matrix. The injected air traverses horizontally and vertically through the soil. Anything that impedes the flow of air will impact the utility of air sparging.   Flow impedance may be caused by lithological barriers that block the vertical flow of the air.   It may also be caused by channelization where the horizontal air flow is "captured" by high permeability channels.   The issue in considering structural limitations to air sparging is understanding barriers to flow.

Second, with respect to operational concerns, it is important to keep in mind that the injection of air can displace both vapors and water. Unless control is **established** and **maintained**, this displacement can accelerate and aggravate the spread of contamination. The issue in considering operational issues is understanding the control of flow.

4.6.   BARRIERS TO FLOW

The effectiveness of air sparging is dependent on the unrestricted flow of air horizontally and vertically through the soil matrix.   Anything that restricts or channels air flow limits air sparging.

Geological barriers will obviously impact air flow. With a sparge system, air flow must be both horizontal and vertical. The vertical travel is important for the ultimate removal of the volatilized contaminant.   If the geology restricts vertical air flow, then sparging can push the dissolved contamination downgradient as shown in Figure 4.6. Any less pervious zone (such as a clay barrier) above the zone of air injection may restrict vertical air flow and severely reduce the effectiveness of air sparging. The barriers do not have to be nonpervious but may simply be a gradation to material with a lower permeability, which can restrict vertical air flow.   The presence or absence of such barriers should be determined during installation of the system and through a pilot test study.

A second potential impact of geology is the presence of soil layers having higher permeability than the sparging zone, which may intercept and channel air flow as shown in Figure 4.7. Such channels are likely when the soil matrix is layered or highly heterogeneous. The greater the degree of heterogeneity, the higher the risk of channeled flow.   Channeled air flow may cause the uncontrolled spread of contamination.   To minimize the risk of channelization, a complete lithological profile of the sparging area should be developed before the system is installed. The importance of channelized flow should be evaluated during a pilot test.

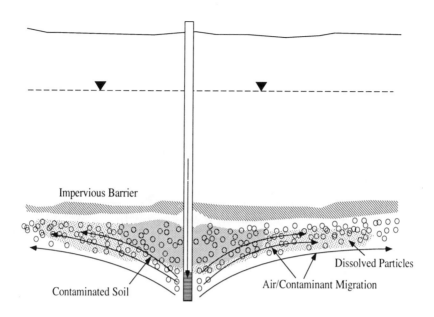

Figure 4.6.   Inhibited vertical air flow due to impervious barrier.

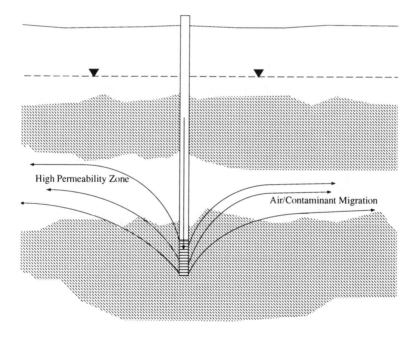

Figure 4.7.   Channeled air flow through highly permeable zone.

4.7. CONTROL OF FLOW

There are two potential concerns with the use of air sparging. Injection of air can displace both vapors and water, and this displacement can accelerate and aggravate the spread of contamination. Therefore, air flow must be controlled during a sparging operation.

There are two operational conditions that may potentially cause a spread of dissolved contaminants. The first is the injection pressure and flow. The second is water table mounding.

A potential cause of dissolved contaminant migration, is over pressurizing the sparge system. As discussed above, too high an air flow rate can physically displace water. An illustration of this is afforded by air rotary drilling. An air rotary rig uses high pressure/volume air (~100-200 psi, 300-500 cfm) to lift and remove cuttings. When drilling below the water table, air rotary rigs have been known to "pop" covers off of adjacent monitoring wells and cause water geysers.

Ostensibly, the minimum injection pressure is that which is required to overcome the water column (i.e., 1 psi for every 2.3 feet of hydraulic head). As pressure is increased above this minimum, air is "injected" laterally into the aquifer. As seen in Figure 4.8, there is an initial linear relationship between the sparge pressure and direction of air travel. At low sparge pressure (injection pressure equal to hydraulic head) the air travels 1-2 feet horizontally for every foot of vertical travel. As the sparge pressure increases, the degree of horizontal travel also increases. Enhanced horizontal travel allows a single well to treat a greater area of aquifer. However, increasing the pressure does not always provide a benefit. Increased pressure may cause air flow to become turbulent, and the added pump energy is wasted. The danger under turbulent conditions is that a dissolved plume of contaminants could be pushed away from the sparge well. Figure 4.8 shows a point of inflection, where the increase in injection pressure does not give a corresponding increase in air flow radius. This transition to turbulent flow is also observed in venting systems where high vacuum can result in frictional heating of the vent gases.

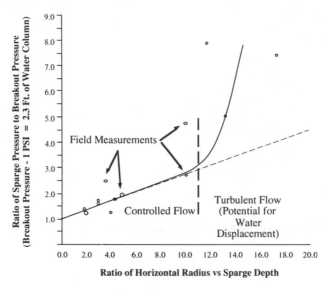

Figure 4.8. Effect of injection pressure on air flow.

A second potential cause of increased dissolved contaminant migration is water table mounding. Air sparging does raise the level of the water table. Normally, ground water would flow away from a mound. However, the mounding produced by sparging is caused by the displacement of water with air. Flow away from the mound may not be induced because the net density of the water column is decreased, thus counteracting the mounding. This lowered density is dramatically seen by taking water table measurements after the sparge system was shut off as shown in Table 4.3.

TABLE 4.3. WATER TABLE MOUNDING AND COLLAPSE

Depth to Water (ft) @

Well # -	Distance from Sparge Point	Static Water Level	Sparging Water Level	5 Min After	10 Min After
MW-7	5	6.46	4.09	10.03	6.96
SE-1919	6.42	6.20	6.93	6.54	
S-2629	6.71	6.55	6.96	6.77	
NE-13	13	6.52	6.11	7.44	6.75

The water table collapses back to its original elevation after air injection is stopped. This collapse shows the displacement of water by air during sparging. Mounding and collapse is greater for monitoring points close to the sparge point. Because of this density compensation, mounding may not spread any contamination.

Additionally, if the air flow rate is high enough during sparging, any dissolved constituents can be stripped before they migrate away from the treatment area. Sparging has been successfully applied with no evidence of ground-water contaminant migration (Brown et al., 1991; Middleton and Hiller, 1990).

The second "danger" of sparging is accelerated vapor travel. This is of concern when the product is volatile and where there are receptors. Since air sparging increases pressure in the vadose zone, any exhausted vapors can be drawn into building basements. Basements are generally low pressure areas, and this can lead to preferential vapor migration and accumulation in basements. As a result, in areas with potential vapor receptors, air sparging should be done with a concurrent vent system. A vent system provides an effective means of capturing sparged gases.

4.8. SUMMARY OF LIMITATIONS

As with any technology, there are limitations to the utility and applicability of air sparging. Understanding those limitations is important to the proper development and use of air sparging. As discussed above, there are several types of limitations to air sparging technology. The first is the type of contaminant. For air sparging to be effective as a removal mechanism, the contaminant must be volatile and insoluble. If the contaminant is soluble or nonvolatile, the contaminant must be biodegradable. In the case of volatile or insoluble contaminants, air sparging functions as an extractive process as well as a biodegradative process. In the case of biodegradable contaminants, air sparging is a destructive process.

The second limitation to air sparging technology is the geological character of the site. The most important geological characteristic is structural homogeneity or heterogeneity. If there is stratification present in the saturated zone, there is a possibility that sparged air could be held below an impervious layer and thus spread laterally, causing the contamination to spread. To guard against such occurrences, any bore hole to be used as a sparge point should be logged by continuous coring over its entire depth before installation of the well. If stratification is present and sparging is to be used, all lithological units above the sparge interval should be of equal or greater permeability compared to the unit in which the sparge point is screened. Optimally the permeability of the geologic material above the screened interval of the sparge well should increase with increasing elevation until the water table is reached.

The second most important geological characteristic is permeability. In many geological environments, the permeability in the vertical direction is less than permeability in the horizontal direction. The permeability should be sufficient to allow the sparge air to move through the aquifer matrix, both horizontally and vertically. If flow is impeded in either direction because of low permeability, then sparging may be precluded. If the ratio of horizontal to vertical permeability is low (<2:1), sparging can be effective even though the general permeability is also low (>10^{-5} cm/sec). If the ratio of horizontal to vertical permeability is high (>3:1), then the general permeability must be higher (>10^{-4} cm/sec) for sparging to be effective.

Finally, there are physical constraints on the operation of a sparge system. The constraints are primarily depth related. There is both a minimum and maximum depth for a sparge system. The minimum depth, 4 feet, is the saturated thickness required to confine the air and force it to "cone-out" from the injection point. If there is insufficient saturated thickness, then the air could short-circuit around the sparge point. The maximum sparge depth, 30 feet, is important from the standpoint of control/predictability. At depths greater than 30 feet it is difficult to predict where the sparge air will travel, making it difficult to design a control system for containing the sparged air with the area being treated and/or to capture the sparge air once it exits the saturated zone. Any small layers with low permeability lying above the sparge interval could have a drastic effect on the air movement. The ability to detect such layers becomes increasingly more difficult with greater sparge depths. Thus the risk of improper design/control also increases. A second depth related constraint is the depth to water. There needs to be sufficient unsaturated soils to allow for the installation of a soil vapor extraction system so that the VOCs mobilized by sparging can be captured. The minimum depth that is required for installation of a soil vapor extraction system is four (4) feet.

To assure the effectiveness of a sparge system, proper consideration must be given to the potential limitations to the technology. As discussed above, these limitations are based on the properties of the contaminant being treated, the geological characteristics of the site, and on the physical limitations to the technology. Table 4.4 summarizes the limitations to sparging.

TABLE 4.4. LIMITS TO THE USE OF AIR SPARGING

FACTOR	PARAMETER	LIMIT / DESIRED RANGE
Contaminant	Volatility	>5 mm Hg
	Solubility	<20,000 mg/l
	Biodegradability	BOD_5 >.01 mg/l, BOD_5:ThOD >.001
Geology	Heterogeneity	No impervious layers above sparge point
		If layering present hydraulic conductivity increases above sparge point
	Hydraulic conductivity	>10^{-5} cm/s if horizontal: vertical is <2:1
		>10^{-4} cm/s if horizontal: vertical is >3:1
Physical	Sparge Depth	>4 feet, <30 feet
	Depth to Water	>4 feet

4.9. SYSTEM APPLICATION AND DESIGN

The best assurance for effective system performance is proper system design. Air sparging systems can only be properly designed through collection of appropriate site data and field pilot testing. Field pilot testing necessitates the installation of test sparge/vent points. The installation of the pilot test system provides an opportunity to identify any subsurface barriers or irregularities that may restrict air flow. An air sparging system can be correctly designed only if sufficient data concerning site conditions have been determined. The data requirements consist of:

1. the nature and extent of site contaminants,
2. specifics of the site hydrogeology, and
3. thorough knowledge of potential ground water and vapor receptors.

4.9.1. Nature and Extent of Site Contaminants

The volatility of the petroleum hydrocarbon being treated should be determined. The higher the volatility, the more vapor transport will be a factor. Vapor transport can be beneficial in that it accelerates treatment by physically removing volatile petroleum hydrocarbons. However, vapor transport can also be a problem in that it must be controlled and treated.

The mass distribution of the site contamination must be known in order to effectively utilize air sparging. The vertical extent of adsorbed phase contaminants at or below the water table must be determined in order to effectively determine the depth of sparging wells. The lateral extent of adsorbed phase contamination below the water table must be known to ensure complete remedial system coverage. In addition, the down-gradient dissolved ground-water concentrations should be delineated in order to allow monitoring of the plume during sparging operation and placement of recovery or sparge wells for ground-water treatment.

4.9.2. Hydrogeologic Conditions

Several hydrogeologic parameters are of great concern in ensuring correct design and operation of an air sparge system. The soil texture must allow for air transmission in order for volatilization or biodegradation to occur. In general a hydraulic conductivity of >10^{-5} cm/sec is necessary for effective air sparging. Poorly compacted fill materials are also a poor choice for an air sparging system as they may exhibit settling if subjected to high air pressures.

Of even more importance is the homogeneity of the site soils. Permeability contrasts due to natural stratigraphic changes or differential filling by human activity will alter the air flow. Lower permeability lenses will create a barrier to the upward moving air and can cause lateral spread of the contaminants. High permeability channels may "capture" the air stream and also cause contaminants to spread.

4.9.3. Potential Ground-water and Vapor Receptors

Since injected air flow can displace both vapors and liquids, the proximity of vapor or ground-water receptors should be determined before a sparging system is installed and operated. If such receptors are present, system safeguards should be used.

The first safeguard is the use of soil vent systems. Soil vent systems are mandatory where volatile hydrocarbons are being treated and there are potential vapor receptors, or where vapor phase controls are required. The vent system should be designed to have a greater flow than the sparge system and should have a greater radius of influence. Barrier soil vent systems can also be placed between the sparge system and potential vapor receptors.

The second safeguard is to utilize ground-water control to prevent the migration of dissolved contaminants. Active pumping systems or effective barriers should be installed. Air flow does not follow hydraulic gradients. Therefore, ground-water control should be installed where receptors exist and not just downgradient. Ground-water controls may be necessary in areas where the geology is heterogeneous or of low hydraulic conductivity (<10^{-4} cm/sec). **Ground-water controls may consist of water collection systems or of an outer sparge system used as a barrier system.**

4.10. FIELD PILOT TESTING

Air sparging requires a balanced air flow. Too low a flow can result in a loss of remedial effectiveness; too high a flow can result in a loss of control. Because of the potential for loss of control, an air sparge system should *never be installed without a pilot test*. Installation of a sparge system requires proper design of the separate components - the vent system (if volatile hydrocarbons are present) and the sparge system, as well as balancing of the two components. The basic design data to be determined by a pilot test are:

1) The radius of influence of the air sparging system conducted at different injection flows/pressures.
2) The radius of influence of the vacuum extraction system.
3) The pressure and vacuum requirements for effective treatment and effective capture of volatilized materials.

The field tests consist of up to three sequential tests. The first test is a sparge radius of influence test. The second is a vacuum radius of influence test. The third is a combined sparge/vent test. The second and third are required at sites where vapor levels are a concern.

A number of different parameters can be measured during the tests to determine radius of influence. These include:

- Vacuum or pressure vs. distances. This is an indication of radius of influence.
- VOC concentrations in soil vapor or ground water. This is an indication of what is being removed and of areas being impacted. Concentrations should be determined before, during (with and without the system running) and after each test.
- CO_2 and O_2 levels in soil vapor. This is an indication of biological activity. These measurements need to be taken before, during and after each pilot test under static as well as pumping conditions.
- Dissolved oxygen (DO) levels in water. This is a good indicator of effectiveness. In areas contaminated with petroleum hydrocarbons, static DO levels are generally < 2 mg/l. Sparging should raise the DO level substantially. Good initial DO measurements are required to determine changes.
- Water levels before and during test. Air flow during sparging will cause some mounding. Levels should be recorded before the test to determine background.

Using multiple parameters allows for cross correlation during design. With this cross correlation, it is possible to determine effective air flow through the area of contamination and ensure capture of the volatilized materials. There is generally good agreement among parameters as shown in Figure 4.9.

4.11. DESIGN DATA REQUIREMENTS

At the conclusion of the site characterization and pilot test, a complete set of design data should have been collected. Table 4.5 lists the different data required and their significance for design.

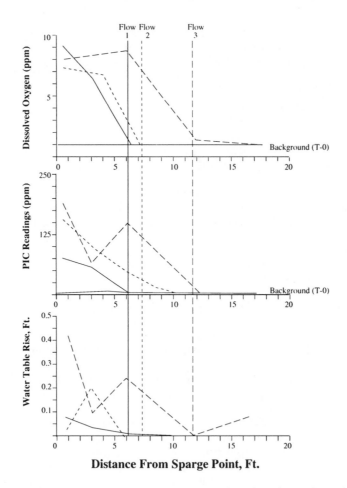

Figure 4.9.   Agreement between sparge parameters in estimating the radius of influence.

TABLE 4.5.   SITE AND PILOT TEST DATA NEEDED FOR DESIGN

DATA	IMPACT ON DESIGN
Lithological Barriers	Feasibility/Sparging Depth
Vertical Extent of Contamination	Sparging Depth
Horizontal Extent of Contamination	Number of Sparge Wells
Volatility of Contaminant	Vapor Control (Venting)
Sparge Radius of Influence	Well Spacing/Flow Requirement
Optimal Flow Rates	Compressor Size
Vent Radius of Influence	Well Spacing
Vacuum/Pressure Balance	Blower Size/Well Placement
Vapor Levels	Vapor Treatment

4.12. SYSTEM ELEMENTS

A sparge system consists of a number of different elements. Some of them are essential, while others are optional. Their use is dictated by the type of product or by the site conditions. Table 4.6 includes a list of the system elements and their importance for sparging.

TABLE 4.6. AIR SPARGING SYSTEM ELEMENTS

COMPONENT	IMPORTANCE	DESCRIPTION
Sparging Well	Essential	Sparging wells consist of a small section of pervious pipe (slotted pipe or diffuser) placed at the bottom of a borehole, set below the zone of contamination (Figure 4.10). A sparge interval should be set every 10-15 ft below the water table. The borehole is grouted above the sparge interval. Above 15 psi, well material should be steel (Compressed Gas Association, 1989)
Air Compressor	Essential	The compressor should be capable of delivering 10-20 cfm per well at 1-3 times breakout pressure. Breakout pressure = (Depth/2.3). Flow rate is determined by ground-water flow and soil volume being treated. An air to water ratio of 10-20:1 is desirable.
Monitoring System	Essential	A monitoring system is necessary to achieve and maintain control. The basic system (Figure 4.11) should consist of a shallow monitoring well to measure water table elevation, DO, VOCs, and pressure as well as vadose zone vapor probes to monitor VOCs and pressure/vacuum.
Heat Exchanger	Optional	A heat exchanger, i.e., thermal vanes, should be used with PVC (Air Inlet) systems, depending on injection pressure.
Muffler	Optional	In urban areas, a muffler may be required to meet noise abatement requirements.
SVE System	Optional	Where volatile hydrocarbons are being treated and vapor receptors or vapor control requirements exist, an SVE system is necessary to capture the VOCs mobilized by the sparge system. The SVE system should be designed such that a net negative pressure is maintained in the treatment area. The total flow of the SVE system should be at least twice the total flow of the sparge system. SVE system elements include wells or trenches and a vacuum pump or blower.
Vapor Treatment	Optional	If air quality is a concern, vapor treatment may be required for the SVE system. Vapor treatment options include thermal treatment (catalytic or thermal oxidation) as well as biotreatment. Petroleum hydrocarbon vapors are generally biodegradable and may be effectively treated by diffusion through a soil bed or compost bed.
Ground-water Control	Optional	If ground-water receptors or control of existing migration are an issue, or if there is a risk of ground-water migration due to high heterogeneity, then a ground-water control system may be required. This may consist of barrier or interceptor wells, or of a barrier sparge system. With petroleum hydrocarbons, dissolved contaminants will be stripped by the air flow and will be biodegraded by enhanced oxygenation. Therefore long term ground-water controls are not an essential part of the system. The primary issue in deciding if ground-water controls are required is the need for immediate containment of the contamination.

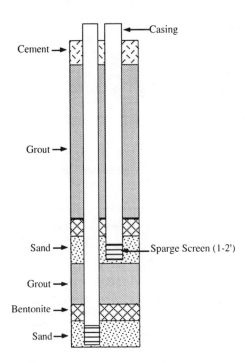

Figure 4.10.   Nested sparge well.

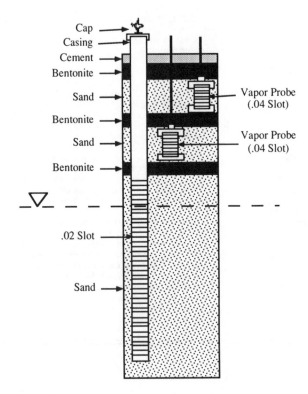

Figure 4.11.   Monitoring point for sparging systems.

4.13. SYSTEM EXAMPLES

The following two examples show the installation of an air sparging system for shallow ground-water aquifers:

Site A:

The subsurface environment at the site generally consists of fill material overlying a continuous sheet of naturally occurring Quaternary sediments (medium sands). Within the southern geographical portion of the property, the Quaternary sediments rest unconformably on top of the sediments of the Potomac Formation, which in turn overlie the basement complex consisting of a volcanic intrusive rock, probably granodiorite. The geology observed during drilling activity indicates that the saturated Quaternary sediments are relatively homogeneous across the property. A natural barrier (clays of the Potomac Formation) exists, which locally minimizes the potential for vertical downward migration of dissolved phase total petroleum hydrocarbon and chlorinated VOCs present in the shallow water-bearing zone into deeper water-bearing units.

The flow of shallow ground water at the property appears to trend northwest (NW) to southeast (SE), under an average hydraulic gradient of 0.021 ft/ft across the property. The gradient does not vary appreciably across the property, ranging from 0.015 ft/ft to 0.026 ft/ft.

The bulk of the contamination appears to be located within two soil horizons; one shallow (~3-9 ft), and one above, at, and just below the water table (15-18+ ft). Thus, it may be concluded that there is soil contamination in both the unsaturated (vadose) and saturated zones (water table aquifer). Second, the soil contamination is primarily isolated to the former tank field area and extends hydrogeologically laterally and downgradient in the direction of shallow ground-water flow. Based on analysis of the soil, it is estimated that approximately 300 to 500 pounds of contamination exist in the upper horizon and an additional 200 to 300 pounds exist in the lower horizon.

Using data obtained during pilot testing, a pattern of vent and sparge points was developed to provide overlapping influence (negative net pressure) and favorable site coverage for the treatment system.   Additional probe nests were strategically placed to monitor system performance. A complete list of treatment and monitoring points installed at the site is specified below, and pictured in Figure 4.12.

- 7 Combination vapor extraction/air sparge points (AS/VP1-AS/VP7); to be installed.
- 1 Vapor extraction only point (VP1).
- 7 Sparge only points (AS1-AS7).
- 8 Vapor monitoring probe nests (PR1-PR8).

The 7 vent/sparge points form a rough ellipse surrounding the former tank field area and extending to the property perimeter. The 7 innermost sparge only points were specified to complete coverage and to provide concentrated treatment within the former tank field area, where contaminant levels are highest.

The vent system was operated at ~40 inches of water vacuum and ~60 CFM per point for a total flow of ~500 CFM. The sparge system was operated at 10 psi and a flow of 16 CFM per point for a total flow of ~225 CFM. Sparge system flow rate was designed to be <one-half of the vent flow.

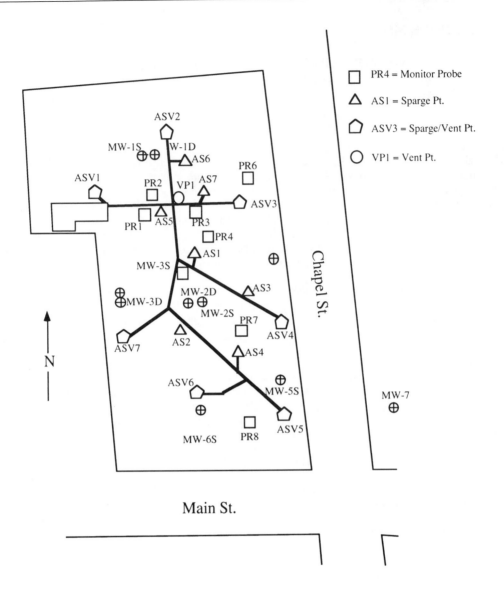

Figure 4.12. **System layout Site A.**

Site B:

Based upon observations made during drilling, the site geology material consists of approximately eight to fifteen feet of a brown to black sandy, silty clay with minor occurrences of ash, gravels and other construction debris. The fill material rests unconformably atop the Marcellus Shale, which is continuous to at least 35 feet below surface grade. The Marcellus Shale exhibits a deteriorated surface at the contact between the unconsolidated and consolidated material, indicating that the surface was previously exposed to weathering processes. On the northern portion of the facility, the shale exhibits minor fracturing, but the shale is competent on the southern portion of the facility.

Ground water occurs within the unconsolidated sediments at a depth ranging from approximately three to six feet below surface grade under water table conditions. Shallow monitoring well elevational data indicate the major component of ground-water flow is approximately north to south. A minor component of ground-water flow appears to occur preferentially northeast to southwest. The hydraulic gradient at the water table is estimated to average 0.015 ft/ft across the property.

Air sparging tests were conducted at three pressure levels; 10, 12 and 14 psi. The test used a 100 psi diesel-powered air compressor. The sparge test points (SP1 and SP2) were installed to a depth of 11 and 11.5 feet, respectively, below the water table and are screened for the bottom two feet. Sparge tests were performed by injecting air into the sparge test point at 10, 12 and 14 psi; the flow rates that correspond to these pressure levels ranged from 40 to 53 CFM. The resulting pressure and VOC levels were measured in the seven vapor probes, which were previously utilized for the vapor extraction tests. Sparging influence was considered present at an induced pressure level of 0.01 inches of water column.

The sparge results appeared to be radial and did not indicate any directional orientation/correlation. Increasing the applied pressure did not appear to have any effect on the radius of influence at a given test point. The change in VOC concentrations during the sparge tests was significant. In most probes the VOC levels, as indicated by OVA readings, increased significantly with sparging. During the sparge tests a 31 to 40 foot radius of influence (at 0.01 inches of water column induced pressure) was developed. This test demonstrated the feasibility of sparging for VOC treatment at and below the water table.

Based on data obtained during the hydrogeologic investigation, soil gas survey and pilot tests, a pattern of vapor extraction trenches and sparge points was developed to provide overlapping influence (negative net pressure) and favorable site coverage for the treatment system. A complete list of treatment and monitoring points for the system is specified below.

- 36 sparge points (to be installed);
- 2,200 feet of vapor extraction trenches (to be installed);
- 12 stainless steel vapor monitoring drive point probes (PR1-PR12 to be installed);
- 11 shallow monitoring wells (MW1-3, 4S, 5-7 existing; MW8-11 to be installed);
- 1 deep monitoring well (MW-4D existing); and,
- 5 potential sump locations.

In order to provide assurance that adequate vacuum would be induced across the site, the pattern of vent locations necessary for full coverage was determined by assuming a maximum trench spacing of 45 feet. This is one-half the calculated radius of influence (ROI) of the vapor extraction trench network. This treatment system layout is designed to maintain a net negative pressure and thus capture VOC contaminated soil gas both on and off the property. Figure 4.13 indicates the proposed location of the vapor extraction trenches and the sparging wells.

Vapor extraction will be accomplished using a 50 Hp blower, having a capacity of 2,500 CFM at 60 inches of water column vacuum. Influent vacuum/flow rate will be controlled with an ambient air intake valve. A liquid knockout tank, particulate filter and muffler will be placed on the influent line to eliminate or reduce water generated during system operation, solids, and noise, respectively. An effluent muffler was specified to further reduce noise levels to minimize the impact to nearby residents. A 12,000 pound granular activated carbon (GAC) unit was specified on the vapor extraction effluent to remove contaminants from the extracted air prior to discharge.

In order to ensure favorable site coverage, a 30 foot radius of influence was assumed in designing the pattern of points to be utilized for the final system. A total of 36 sparge locations were specified to provide this coverage (Figure 4.13). The proposed location of the sparge points (to the east and south of the facility building) was based on ground water and soil analytical results obtained during the hydrogeologic investigation. Operation of 36 sparge points at 40 CFM each results in a potential total flow rate of 1,440 CFM. However, as with vapor extraction, significant competition between points will result in reduced air flow rates. Based on this fact and experience from a similar site, a flow rate of 860 CFM was used for the design. The vapor extraction system should be able to easily capture the sparge air since the sparge system design flow rate is roughly one-third of the vapor extraction system design flow rate of 2,500 CFM. The sparge air will be provided by a 75 Hp rotary lobe-type blower capable of delivering 860 CFM at 10 psi.

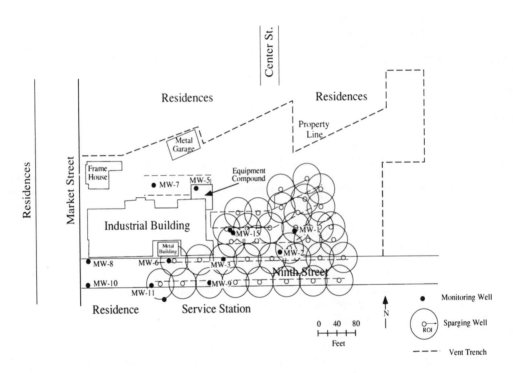

Figure 4.13.   Layout of site B air sparging/vent system.

4.14.   COST FACTORS

The cost for a sparge system is dependent on the size of the site, the degree of contamination, the application depth, the geology (permeability, heterogeneity) of the site, and the permitting requirements. Table 4.7 lists the approximate costs for a one acre site having shallow ground water (DTW <20 ft), and a moderate permeability (fine to medium sand).

TABLE 4.7. APPROXIMATE COST FACTORS

FACTOR	COST, $	OPTIONAL	COST INFLUENCES
Sparge Well (PVC)	3,000	N	Depth, Diameter
Sparge Well (Steel)	5,000	Y	Depth, Diameter
Total Wells (10)	30,000	N	ROI[a], Area
Vent Wells (5 additional)	12,000	Y	ROI[a], Sparge flow, Area, DTW[b]
Piping & Trenching (PVC)	45,000	N	Area, Depth
Piping & Trenching (Steel)	60,000	Y	Area, Depth
Compressor (300 cfm, 20 psi)	15,000	N	Flow, Pressure
Vacuum Blower (1000 cfm, 100")	20,000	Y	Flow, Vacuum
Vapor Treatment (Carbon)	180,000	Y	Flow, Concentration
Vapor Treatment (Thermal)	120,000	Y	Flow, Concentration
Permitting (Air)	15,000	Y	Regulations
Permitting (Water)	25,000	Y	Regulations
Pilot Test	20,000	N	Depth, Geology
Design	25,000	N	Flow, VOCs, Area, Regulations
Construction	50,000	N	Flow, VOCs, Area, Regulations
O&M (per Year)	75,000	N	Regulations, Flow, VOC Concentrations
Reporting (per Year)	30,000	Y	Regulations
Total without Options	260,000		
Total with Vapor Control	432,000		
Total Maximum	570,000		

[a] Radius of influence
[b] Depth to water

4.15. CONCLUSION

While soil vapor extraction has long been recognized as an effective means of removing volatile organics from subsurface soils, it has been limited to treatment of unsaturated soils. Where contamination exists below the water table, soil vapor extraction is limited and can only be used with an extensive and often costly dewatering operation.

Air sparging is a means of extending the utility of vapor extraction technology to the saturated regime. With air sparging, air is injected under pressure below the water table, creating a transient air-filled porosity. This enhances biodegradation as well as volatilization of petroleum hydrocarbon contaminants from the soil and ground water. The net result is a rapid and significant decrease in contaminant levels.

Air sparging has two inherent dangers. First, the VOC-laden air stream can rapidly migrate through the vadose zone to low pressure zones such as basements, causing a vapor hazard. To prevent this occurrence, a sparge system should be operated in conjunction with a vent system. A second danger is that the injected air can mobilize ground-water contaminants rather than stripping them, causing accelerated downgradient migration. This can occur if there are vertical barriers to air migration causing the air to be trapped producing lateral spread. It can also occur if too much pressure is used, physically displacing the water column.

Because of these dangers, proper design is essential. This necessitates a field pilot test and careful site delineation. With proper design, the use of sparging can substantially and rapidly remediate ground-water contamination. The key to the effective use of air sparging is proper design. This entails first understanding the distribution of contaminants across the site. Second, the flow dynamics of air must be determined in the vadose zone and in the saturated zone. This can only be done through a properly designed pilot study. Once both the contaminant distribution and the flow dynamics are known, the number, location and type of treatment wells can be specified. This, in turn, leads to the equipment specifications. Through a careful and phased design process, air sparging can be an effective remedial system.

To further expand the utility of sparging, there are a number of questions that need to be addressed. These questions are:

- What are the limitations to air sparging technology?
- How does air sparging impact the site hydrogeology and contaminant transport?
- What are the most effective means of determining the radius of influence, pressure requirements, and effectiveness of a sparge system to minimize detrimental effects?

With effective design and careful monitoring, air sparging can be an important remedial tool. If it is applied in a simplistic fashion, air sparging can be ineffectual at best or counter-productive at worse.

REFERENCES

Brown, R.A., C. Herman, and E. Henry. 1991. The use of aeration in environmental clean-ups. In: *Proceedings, Haztech International Pittsburgh Waste Conference.* Pittsburgh, Pennsylvania. May 1991.

Clayton, W.S., R.A. Brown, and K.L. Brody. 1989. The reduction of groundwater contamination by vapor extraction of volatile organics from the vadose zone. In: *New England Environmental Expo.* Boston, Massachusetts. May 1989.

Compressed Gas Association. 1989. *Handbook of Compressed Gases.* Van Nostrand Reinhold. New York, New York. 657 p.

Hiller, D., and H. Gudemann. 1988. In situ remediation of VOC contaminated soil and groundwater by vapor extraction and groundwater aeration. In: *Proceedings, Haztech '88 International.* Cleveland, Ohio. September 1988.

Hoag, G.E., and M.C. Marley. 1984. Induced soil venting for the recovery/restoration of gasoline hydrocarbons from the vadose zone. In: *Proceedings of the Petroleum Hydrocarbons and Organic Chemicals in Ground Water Conference.* National Water Well Association, American Petroleum Institute. Houston, Texas. November 1984.

Marley, M.C., M.T. Walsh, and P.E. Nangeroni. 1990. Case study on the application of air sparging as a complementary technology to vapor extraction at a gasoline spill site in Rhode Island. In: *Proceedings, HMC Great Lakes 90.* Hazardous Materials Control Research Institute. Silver Spring, Maryland.

Middleton, A.C., and D.H. Hiller. 1990. In situ aeration of ground water - a technology overview. In: *Proceedings, Conference on Prevention and Treatment of Soil and Groundwater Contamination in the Petroleum Refining and Distribution Industry.* Montreal, Quebec, Canada. October 1990.

Raymond, R.L., V.W. Jamison, and J.O. Hudson. 1975. Biodegradation of high-octane gasoline in groundwater. *Development in Industrial Microbiology.* Volume 16. American Institute of Biological Sciences. Washington, DC.

SECTION 5

GROUND-WATER TREATMENT FOR CHLORINATED SOLVENTS

Perry L. McCarty
Lewis Semprini
Western Region Hazardous Substance Research Center
Stanford University, Stanford, California 94305-4020
Telephone: (415)723-4131
Fax: (415)725-8662

5.1. INTRODUCTION

Chlorinated solvents and their natural transformation products represent the most prevalent organic ground-water contaminants in the country. These solvents, consisting primarily of chlorinated aliphatic hydrocarbons (CAHs), have been used widely for degreasing of aircraft engines, automobile parts, electronic components, and clothing. Once dirty, chlorinated solvents often have been disposed into refuse sites, waste pits and lagoons, and storage tanks. Because of their relative solubility in water and their somewhat poor sorption to soils, they tend to migrate downward through soils, contaminating water with which they come into contact. Being denser than water, their downward movement is not impeded when they reach the water table, and so they can penetrate deeply beneath the water table. CAHs have water solubilities in the range of 1 g/l, or several orders of magnitude higher than the drinking water standards for those that are regulated.

The major chlorinated solvents used in the past are carbon tetrachloride (CT), tetrachloroethene (PCE), trichloroethene (TCE), and 1,1,1-trichloroethane (TCA). These compounds can be transformed by chemical and biological processes in soils to form a variety of other CAHs, including chloroform (CF), methylene chloride (MC), *cis*- and *trans*-1,2-dichloroethene (*cis*-DCE, *trans*-DCE), 1,1-dichloroethene (1,1-DCE), vinyl chloride (VC), 1,1-dichloroethane (DCA), and chloroethane (CA). These chemicals, their solubilities in water, and drinking water maximum contaminant limits (MCL), if applicable, are listed in Table 5.1. This is the group of chemicals generally to be addressed as a result of chlorinated solvent contamination of ground water.

Just over one decade ago, most of the compounds listed in Table 5.1 were considered to be nonbiodegradable. Transformation products of the chlorinated solvents then started to be found in ground waters, and this led to expanded efforts to determine the chemical and biological processes responsible. It was found that most of the CAHs can in fact be transformed by biological processes, but generally, the microorganisms responsible cannot obtain energy for growth from the transformations. The transformations are brought about by cometabolism, or through interactions of the CAHs with enzymes or cofactors produced by the microorganisms for other purposes. There are now widespread efforts to take advantage of cometabolism for the transformation of CAHs in ground water, but this is a much more complicated process than the usual biological treatment processes that have been used for years, in which organic compound destruction is accomplished by organisms that use the compounds as primary substrates for energy and growth. In cometabolism, other chemicals must be present to serve as primary substrates to satisfy the energy needs of the microorganisms, and indeed must be tailored so that they can stimulate the production of the biological agents that affect cometabolism of the CAHs.

TABLE 5.1. COMMON HALOGENATED ALIPHATIC HYDROCARBONS

Compound	Formula	Acronym	Density	Water Solubility (mg/l)	U.S. Drinking Water MCL (μg/l)
Carbon Tetrachloride	CCl_4	CT	1.595	800	5
Chloroform	$CHCl_3$	CF	1.485	8,200	100
Methylene Chloride	CH_2Cl_2	MC	1.325	13,000	
1,1,1-Trichloroethane	CH_3CCl_3	TCA	1.325	950	200
1,1-Dichloroethane	CH_3CHCl_2	1,1-DCA	1.175	5,500	
1,2-Dichloroethane	CH_2ClCH_2Cl	1,2-DCA	1.253	8,700	5
Chloroethane	CH_3CH_2Cl	CA			
Tetrachloroethene	$CCl_2=CCl_2$	PCE	1.625	150	5
Trichloroethene	$CHCl=CCl_2$	TCE	1.462	1,000	5
cis-1,2-Dichloroethene	$CHCl=CHCl$	*cis*-DCE	1.214	400	70
trans-1,2-Dichloroethene	$CHCl=CHCl$	*trans*-DCE	1.214	400	100
1,1-Dichloroethene	$CH_2=CCl_2$	1,1-DCE			7
Vinyl chloride	$CH_2=CHCl$	VC			2

Much has already been learned about cometabolism of CAHs. However, full-scale field applications of this process are greatly limited, and there are virtually no sufficiently well-documented full-scale applications at present that can be used to guide design and application or that can be used to evaluate costs. Thus, any application of bioremediation for chlorinated solvent destruction in the field must be considered as a research activity and should be evaluated as such. As with any new and untested process, failure to reach desired goals should be anticipated, and surprises can be expected. Nevertheless, the understanding of the process is now at a stage where full-scale experimentation is desirable, and indeed is a necessity if biodegradation of chlorinated solvents is to become a reality rather than just a laboratory curiosity.

The purpose of this chapter is to provide background information on the state of knowledge of CAH biodegradation, to discuss field as well as laboratory testing of the process, to summarize the potential application of biological destruction for various CAHs, and to discuss the effect of site conditions on the probability for success of in-situ field applications.

5.2. BIOTRANSFORMATION OF CAHS

5.2.1. Primary Substrates and Cometabolism

Organic compounds can be biotransformed by microorganisms through two basically different processes: (1) use as a primary substrate, and (2) cometabolism. In the first process biodegradation occurs when the organism consumes the organic compound as a primary substrate to satisfy its energy and organic carbon needs. This is the usual process for organic decomposition in nature and the process generally captured in the vast majority of biological treatment processes designed for municipal and industrial wastewaters. Knowledge about organism growth and kinetics of primary substrate utilization is quite extensive.

Cometabolism, on the other hand, is the fortuitous transformation of an organic compound by enzymes or cofactors produced by organisms for other purposes. Here, the organisms obtain no obvious or direct benefit from the transformation. Indeed, it may be harmful to them. Cometabolism is also a natural process, but it has not been used extensively for treatment of organic wastes, and knowledge of the process and its practical application are by comparison quite limited. Cometabolism, however, is the process by which most of the CAHs can be biotransformed. Because of the potential usefulness of biotransformation of CAHs, there is much ongoing research to learn how cometabolism might be applied. For cometabolism to occur, an active population of microorganisms having the cometabolizing enzymes or cofactors must be present. This means that the appropriate primary substrates for growth and maintenance of these organisms must also be present. This is an aspect that adds greater complexity and cost to cometabolic biotransformations.

Biotransformation through primary substrate utilization or through cometabolism may occur under either aerobic or anaerobic conditions. Table 5.2 contains a summary of the CAHs that have been shown to be degraded by the two different processes under aerobic or anaerobic conditions. Relative information on transformation rates under the different processes is also indicated, and transformation rates are discussed subsequently in greater detail.

Transformations of CAHs in the natural environment also can occur both chemically (abiotic) and biologically (biotic). The major abiotic and biotic transformation processes occurring in natural systems are summarized in Table 5.3 (Vogel et al., 1987). The abiotic processes most frequently occurring under either aerobic or anaerobic conditions are hydrolysis and dehydrohalogenation. Abiotic transformations generally result in only a partial transformation of a compound and may lead to the formation of a new compound that is either more readily or less readily biodegraded by microorganisms. Biotic transformation products are different under aerobic than anaerobic conditions. When used as a primary substrate, organic chemicals are generally completely mineralized under both aerobic and anaerobic conditions. However, with cometabolism, as with abiotic transformations, CAHs are generally transformed only partially by the biological process.   The eventual fate depends upon other abiotic or

TABLE 5.2.   POTENTIAL FOR CAH BIOTRANSFORMATION AS A PRIMARY SUBSTRATE OR THROUGH COMETABOLISM

	Primary Substrate		Cometabolism		
Compound	*Aerobic Potential*	*Anaerobic Potential*	*Aerobic Potential[a]*	*Anaerobic Potential[a]*	*CAH Product*
CCl_4			0	XXXX	$CHCl_3$
$CHCl_3$			X	XX	CH_2Cl_2
CH_2Cl_2	Yes	Yes	XXX		
CH_3CCl_3			X	XXXX	CH_3CHCl_2
CH_3CHCl_2			X	XX	CH_3CH_2Cl
CH_2ClCH_2Cl	Yes		X	X	CH_3CH_2Cl
CH_3CH_2Cl	Yes		XX	b	
$CCl_2=CCl_2$			0	XXX	$CHCl=CCl_2$
$CHCl=CCl_2$			XX	XXX	$CHCl=CHCl$
$CHCl=CHCl$			XXX	XX	$CH_2=CHCl$
$CH_2=CCl_2$			X	XX	$CH_2=CHCl$
$CH_2=CHCl$	Yes		XXXX	X	

[a]   0 - very small if any potential; X - some potential; XX - fair potential; XXX - good potential; XXXX - excellent potential.
[b]   Readily hydrolyzed abiotically, with half-life on order of one month.

biotic reactions that might occur. Aerobic biotic transformations generally are oxidations and are classified as hydroxylation, or the substitution of a hydroxyl group on the molecule, or epoxidation, in the case of unsaturated CAHs. The anaerobic biotic processes generally are reductions that involve either hydrogenolysis, the substitution of a hydrogen atom for chlorine on the molecule, or dihaloelimination, where two adjacent chlorine atoms are removed, leaving a double bond between the respective carbon atoms.

TABLE 5.3. TRANSFORMATIONS OF CAHS (AFTER VOGEL ET AL., 1987)

Reactions	*Examples*
I. Substitution	
a. solvolysis, hydrolysis	
$RX + H_2O \rightarrow ROH + HX$	$CH_3CH_2CH_2Br + H_2O \rightarrow CH_3CH_2CH_2OH + HBr$
b. other nucleophilic reactions	
$RX + N^- \rightarrow RN + X^-$	$CH_3CH_2Br + HS^- \rightarrow CH_3CH_2SH + BR^-$
II. Oxidation	
a. α - hydroxylation	
$-\overset{\mid}{\underset{H}{C}}-X + H_2O \rightarrow -\overset{\mid}{\underset{OH}{C}}-X + 2H^+ + 2e^-$	$CH_3CHCl_2 + H_2O \rightarrow CH_3 CCl_2 OH + 2H^+ + 2e^-$
b. epoxidation	
$\underset{/}{\overset{\backslash}{C}} = \underset{\backslash}{\overset{/}{C}} + H_2O \rightarrow \underset{\backslash}{\overset{\backslash}{C}} - \underset{\backslash}{\overset{/X}{C}} + 2H^+ + 2e^-$	$CHClCCl_2 + H_2O \rightarrow CHClOCCl_2 + 2H^+ + 2e^-$
III. Reduction	
a. hydrogenolysis	
$RX + H^+ + 2e^- \rightarrow RH + X^-$	$CCl_4 + H^+ + 2e^- \rightarrow CHCl_3 + Cl^-$
b. dihaloelimination	
$-\overset{\mid}{\underset{X}{C}}-\overset{\mid}{\underset{X}{C}}- + 2e^- \rightarrow \underset{/}{\overset{\backslash}{C}} = \underset{\backslash}{\overset{/}{C}} + 2X^-$	$CCl_3CCl_3 + 2e^- \rightarrow CCl_2 CCl_2 + 2Cl^-$
c. coupling	
$2 RX + 2e^- \rightarrow R - R + 2X^-$	$2 CCl_4 + 2e^- \rightarrow CCl_3 CCl_3 + 2Cl^-$
IV. Dehydrohalogenation	
$-\overset{\mid}{\underset{X}{C}}-\overset{\mid}{\underset{H}{C}}- \rightarrow \underset{/}{\overset{\backslash}{C}} = \underset{\backslash}{\overset{/}{C}} + HX$	$CCl_3CH_3 \rightarrow CCl_2 CH_2 + HCl$

5.2.2. CAH Usage as Primary Substrates

Few of the CAHs have been shown to serve as primary substrates for energy and growth by some microorganisms. Among the C_1 compounds, dichloromethane (DM) can be used as a primary substrate under both aerobic and anaerobic conditions. DM can be completely mineralized while serving as a primary substrate under anaerobic conditions by municipal digesting sludge microorganisms (Rittmann and McCarty, 1980a,b; Klecka, 1982; Freedman and Gossett, 1991). Pure cultures of the genera *Pseudomonas* and *Hyphomicrobium* have been isolated that can grow aerobically on DM as the sole carbon and energy source (Brunner and Leisinger, 1978; Brunner et al., 1980; Stucki et al., 1981; La Pat-Polasko et al., 1984; Kohler-Staub and Leisinger, 1985).

The two-carbon saturated CAH, 1,2-dichloroethane (1,2-DCA), can also be used as a primary energy source under aerobic conditions (Stucki et al., 1983, Janssen et al., 1985). One unsaturated two-carbon CAH, VC, also has been shown to be available as a primary substrate for energy and growth under aerobic conditions (Hartmans et al., 1985; Hartmans and de Bont, 1992). These few exceptions noted to date indicate that only the less halogenated one- and two-carbon CAHs might be used as primary substrates for energy and growth, and that the organisms that are capable of doing this are not necessarily widespread in the environment. The biological transformation of most of the CAHs depends upon cometabolism.

5.2.3. Anaerobic Cometabolic Transformation of CAHs

In 1981, the potential for anaerobic biological cometabolism of brominated and chlorinated (halogenated) aliphatic hydrocarbons was demonstrated (Bouwer et al., 1981). Subsequently, CAHs, in general, have been found to transform under a variety of environmental conditions in the absence of oxygen. Figure 5.1 illustrates the various anaerobic biotic and abiotic pathways that chlorinated aliphatic compounds may undergo at contaminated sites. For example, the chlorinated solvent TCA may be transformed

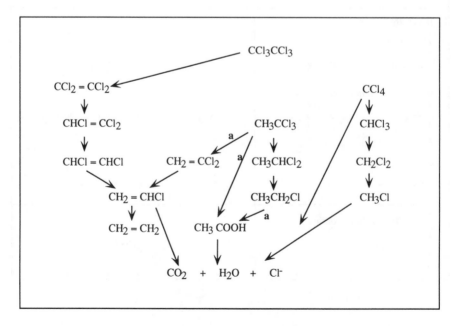

Figure 5.1.   Anaerobic Transformations of CAHs (after Vogel et al., 1987).

abiotically to form 1,1-DCE and acetic acid. The rates are relatively slow, with a half-life for TCA on the order of one year (Vogel et al., 1987; Cline and Delfino, 1989; Jeffers et al., 1989). Also, under anaerobic conditions, TCA may be biologically transformed into 1,1-DCA, which can be further reduced to CA. CA is relatively stable biologically, but abiotically can be transformed into ethanol and chloride, thus rendering it relatively nontoxic. Thus, when TCA is discharged to soil, a variety of abiotic and biotic transformation products may be found there in later years. As another example, TCE can be reduced anaerobically to either *cis-* or *trans-*DCE, both of which can be further transformed into VC. Recent research has indicated that VC can even undergo reduction into ethylene (Freedman and Gossett, 1989; DiStefano et al., 1991), which is essentially harmless.

The pathways outlined in Figure 5.1 suggest that it is possible to render harmless essentially any chlorinated aliphatic compound under anaerobic conditions. While this is true, there are several problems that hinder this potential approach to bioremediation. First, the biotic transformations generally involve cometabolism such that other organic compounds must be present to serve as primary substrates for organism growth. Second, the rates of anaerobic transformation are much greater for the highly chlorinated compounds than for the less-chlorinated compounds, so that the less-chlorinated ones persist longer in the environment. Third, some of the anaerobic transformation products are more hazardous than the parent compounds. Examples here are TCE transformation to VC and TCA transformation to 1,1-DCE. Fourth, reaction rates tend to be greater under highly reducing conditions associated with methane formation than under the less reducing conditions associated with denitrification (Bouwer and Wright, 1988). The latter is the main anaerobic process occurring when excess nitrates are present. Reductive transformation rates are somewhat intermediate between the two under conditions favoring sulfate reduction (sulfate, but no nitrate present). Fifth, when proper environmental conditions are present, microorganisms that can bring about the transformations through cometabolism must also be present. Thus, with anaerobic conditions, one cannot count upon sufficiently high rates and complete transformation to harmless products to occur in ground water unless all the right conditions are present. On the other hand, anaerobic transformation processes do frequently occur, converting chlorinated aliphatic compounds into less chlorinated species that are more readily transformed by aerobic microorganisms. It is for this reason, as well as to help understand the environmental fate of compounds, that knowledge of anaerobic pathways is important.

5.2.4. Aerobic Microbial Transformation of Chlorinated Aliphatic Hydrocarbons

Although some CAHs, particularly those with few chlorines on the molecule, were shown to be biodegradable by microorganisms some time earlier, knowledge that a broader range of CAHs can be oxidized aerobically through cometabolism is rather recent. Wilson and Wilson (1985) showed for the first time that TCE may be susceptible to aerobic degradation through use of soil microbial communities fed natural gas. The processes involved are illustrated by the following equations for TCE cometabolism by methanotrophic bacteria, organisms that oxidize methane for energy and growth:

Methane Oxidation:

$$CH_4 \xrightarrow[\text{NADH, } O_2]{\text{MMO}} CH_3OH \longrightarrow H_2CO \longrightarrow HCOOH \longrightarrow CO_2 \qquad (1)$$

Synthesis   NADH   NADH

TCE Epoxidation:

$$CCl_2 = CHCl \xrightarrow[\text{NADH, O}_2]{\text{MMO}} Cl_2 C \overset{O}{\triangle} CHCl \longrightarrow \longrightarrow CO_2, Cl^-, H_2O$$

(2)

Methanotrophs use an oxygenase (methane monooxygenase or MMO) to catalyze the oxidation of methane to methanol. This requires energy or reducing power in the form of NADH. MMO also oxidizes TCE fortuitously to form TCE epoxide (Little et al., 1988; Fox et al., 1990), an unstable compound that chemically undergoes decomposition to yield a variety of products, including carbon monoxide, formic acid, glyoxylic acid, and a range of chlorinated acids (Miller and Guengerich, 1982). In mixed cultures, as occurs in nature, cooperation between the TCE oxidizers and other bacteria occurs, and TCE is further mineralized to carbon dioxide, water, and chloride (Fogel et al., 1986; Henson et al., 1989; Roberts et al., 1989; Henry and Grbic-Galic, 1991a).

Since the report of Wilson and Wilson (1985) with TCE cometabolism, much scientific research addressing this phenomenon has been performed. The groups of aerobic bacteria currently recognized as being capable of transforming TCE and other CAHs through cometabolism comprise not only the methane oxidizers (Fogel et al., 1986; Little et al., 1988; Mayer et al., 1988; Oldenhuis et al., 1989; Tsien et al., 1989; Henry and Grbic-Galic, 1990; Alvarez-Cohen and McCarty, 1991a,b; Henry and Grbic-Galic, 1991a,b; Lanzarone and McCarty, 1990; Oldenhuis et al., 1991), but also propane oxidizers (Wackett et al., 1989), ethylene oxidizers (Henry, 1991), toluene, phenol, or cresol oxidizers (Nelson et al., 1986, 1987, 1988; Wackett and Gibson, 1988; Folsom et al., 1990; Harker and Kim, 1990), ammonia oxidizers (Arciero et al., 1989; Vannelli et al., 1990), isoprene oxidizers (Ewers et al., 1991), and vinyl chloride oxidizers (Hartmans and de Bont, 1992). These microorganisms all have catabolic oxygenases that catalyze the initial step in oxidation of their respective primary or growth substrates and have potential for initiating the oxidation of CAHs. There is currently insufficient information on the relative advantages and disadvantages of the different oxygenase systems to recommend definitively one over the other, but each may have its place. Most research to date has been conducted with the methane oxidizers and the group of bacteria containing toluene oxygenase, which can be induced with primary substrates such as toluene, phenol, and cresol.

The oxygenases for the above organisms are often nonspecific and fortuitously initiate oxidation of a variety of compounds including most of the CAHs. The exceptions are highly chlorinated CAHs such as CT and PCE. In general, oxygenases act on unsaturated CAHs such as TCE, by adding oxygen across the double bond to form an epoxide. With saturated CAHs such as CF or TCA, a hydroxyl group is generally substituted for one of the hydrogen atoms in the CAH molecule. Frequently, the resulting products from CAH oxidation are chemically unstable and decompose as described above for TCE, yielding products that are further metabolized by other microorganisms present in nature.

5.3. PROCESSES AFFECTING CHEMICAL MOVEMENT AND FATE

In order to apply in-situ bioremediation of CAHs, an understanding of factors affecting the movement and fate of contaminants in ground water is needed. Once percolated from the land surface to ground water, organic contaminants such as the chlorinated solvents and petroleum hydrocarbons are subject to a variety of influences that lead to a complex pattern of behavior. The major processes influencing the transport, distribution, and fate of these chemicals in ground water include the following (McCarty et al., 1992):

1. Advection: the miscible transport in aqueous solution under the influence of the hydraulic potential gradient;

2. Dispersion: the mixing and spreading of concentration fronts that arise largely from differential rates of movement along the myriad individual flow paths through the porous medium;

3. Sorption: the partitioning of a compound between the moving solution and the stationary solid phase;

4. Transformation: the result of chemical reactions or microbial activity that may convert an organic compound into stable products or into another intermediate product;

5. Immiscible transport: the migration of slightly soluble chemicals as a separate liquid phase, often driven downward by density difference in the case of chlorinated solvents;

6. Diffusional transport: the slow migration of solute molecules into the matrix rock or dead-end pores under the influence of a concentration driving force.

The influence of these factors on contaminant behavior has been summarized in several reviews (NAS, 1984; Mackay et al., 1985; Goltz and Roberts, 1986, 1987). The following brief discussion focuses on the principles underlying the sorption and transformation processes, with CAHs being used for illustration.

5.3.1. Effect of Sorption

CAHs do not tend to sorb to soils and aquifer materials as readily as do many hazardous chemicals such as pesticides, PAHs, and PCBs. Nevertheless, sorption in aquifer systems is sufficient to retard the rate at which they move in ground water in relation to the movement of ground water itself. This relative movement can be expressed mathematically by the retardation equation (Freeze and Cherry, 1979):

$$v/v_c = 1 + \rho_b K_d/n \tag{3}$$

where,

v = average linear velocity of ground water
v_c = average linear velocity of the contaminant
ρ_b = bulk mass density of solids in aquifer
n = porosity
K_d = distribution coefficient

The term $(1 + \rho_b K_d/n)$ is commonly known as the retardation factor. For aquifer materials, ρ_b is approximately 1.8 g/cm³, and n generally varies between 0.2 and 0.4 (Freeze and Cherry, 1979). With these units, K_d is expressed in units of cm³/g. K_d is defined as follows:

$$K_d = \frac{\text{Contaminant mass on solid phase per unit mass solid phase}}{\text{Concentration of contaminant in solution}} \qquad (4)$$

Roberts et al. (1980, 1982) measured the retardation of various CAHs including the common chlorinated solvents that were present in reclaimed municipal wastewater that was injected into a confined aquifer in Palo Alto, California. The differences in sorption affinity (i.e., K_d) cause differences in mobility that are reflected as a chromatographic separation of the concentration fronts arriving at an observation well. The retardation factors, estimated as the ratio of the center of mass for the respective concentration response relative to that of chloride, were as follows (Roberts et al., 1980): chloroform (CHCl₃), 3; bromoform (CHBr₃), 6; chlorobenzene (C₆H₅Cl), 33, and 1,1,1-trichloroethane (CH₃CCl₃), 12.

From results of a field experiment at the Borden Air Force Base, Ontario, Canada, sorption and other processes affecting contaminant movement can be readily seen (Figure 5.2). Here, a 12 m³ plug of water contaminated with five organic chemicals and a chloride tracer was injected into an unconfined relatively uniform aquifer with small-scale horizontal bedding (Mackay et al., 1986; Roberts et al., 1986). Figure 5.2 illustrates only the movement of CT and PCE relative to chloride as the other three organic chemicals were degraded over the two-year period. The marked chromatographic separation of these CAHs can be seen. Also to be seen in Figure 5.2 is the impact of dispersion on contaminant movement, as indicated by the significant elongation in the longitudinal direction, or direction of flow, while spreading in the lateral and vertical directions was relatively small.

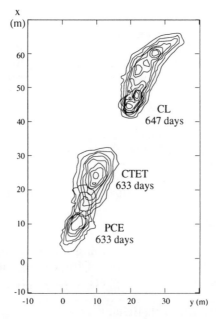

Figure 5.2. Effect of advection, dispersion, and sorption on contaminant movement from an experiment at Borden Air Force Base (after Roberts et al., 1986).

The aquifer material at Borden had a low organic content (0.02%). Measured values of K_d for CT and PCE were 15×10^{-8} m³/g and 45×10^{-8} m³/g, respectively (Curtis et al., 1986a). These values corresponded to retardation factors of 1.9 and 3.6. Although retardation factors inferred from short-term field observations (10-30 days) were consistent with the laboratory-measured K_d values, the study also showed that the retardation factors for these two compounds increased with time and distance from the point of injection to high values of 2.5 for CT and 5.9 for PCE. This appears to have been related partially to a slow rate of diffusion of the contaminants into the aquifer solids, the characteristic time scale which can be measured in terms of weeks to months, rather than hours as commonly assumed. This suggests that short-term laboratory evaluations are not adequate for determining retardation factors in the field. Another factor probably causing the increased retardation with time is aquifer heterogeneities.

Some have indicated that a good correlation exists between K_d and the aquifer organic content and the contaminant's octanol/water partition coefficient, K_{ow} (Karickhoff et al., 1979; Schwarzenbach and Westall, 1981). However, this correlation appears relatively poor for aquifers with low organic content ($f_{oc} < 0.1\%$) (McCarty et al., 1981). Generally, aquifers are fairly poor in organic matter content so that the retardation noted appears to be more a function of sorption to inorganic rather than organic materials (Curtis et al., 1986b).

Retardation is an important process in ground waters for at least two reasons. First, since chemicals have different sorptive properties, their relative rates of movement through aquifers will differ widely (Roberts et al., 1982). Thus, if an aquifer is contaminated with several compounds at one location, each contaminant will move at a different speed, and they will arrive at a downgradient well at different times. The other aspect of importance is that knowledge of the retardation factor provides a basis for estimating the relative amount of the contaminant present in the aqueous phase as compared with that sorbed to the aquifer solids. For example, for a retardation factor of 5, one-fifth of the contaminant is present in the aqueous phase and four-fifths is sorbed onto the aquifer solids. Restoration of a contaminated aquifer requires that the contaminant be removed from the solid phase as well as from the liquid phase. In addition, sorption also tends to reduce the contaminant transformation rate by making the contaminant inaccessible to microorganisms. The effects of these different factors on in-situ bioremediation are discussed later in the section on nutrient introduction and mixing.

5.4. FIELD PILOT STUDIES OF CAH TRANSFORMATION

There is no well-documented full-scale experience with in-situ bioremediation of CAHs upon which to base full-scale application. However, limited small field-scale pilot studies have been conducted in order to determine the effectiveness of certain approaches to remediation. Three efforts have been conducted using the Moffett Naval Air Base pilot facility in Mountain View, California, to evaluate the capacity of native microorganisms (i.e., bacteria indigenous to the ground-water zone) to aerobically and anaerobically cometabolically degrade CAHs when proper conditions were provided to enhance bacterial growth.

The Moffett Field studies were conducted in a shallow confined aquifer under conditions typical of ground-water contamination by CAHs. Two aerobic systems with oxygen as the electron acceptor have been tested: (1) methanotrophs that use methane as a primary substrate and cometabolically transform CAHs with methane monooxygenase

(MMO); and (2) phenol-utilizers in which toluene oxygenase (TO) serves as the cometabolizing enzyme. Target contaminants in these studies were TCE, *cis*-DCE, *trans*-DCE, and VC, in the concentration range of 40 to 120 µg/l. The third study conducted at Moffett Field was an anaerobic, or anoxic, study in which acetate was the primary substrate applied to obtain transformation of CT under denitrification conditions.

The experimental approach taken was similar to that proposed for bioremediation in the field (see Section 5.5). Extracted ground water from the treatment zone was amended with the growth substrate and oxygen, and reinjected to stimulate indigenous growth. CAHs in the extracted ground water were reinjected into the biostimulated zone. Bioremediation conducted in this manner promoted the degradation of inplace contaminants as well as contaminants that were extracted and reinjected, thus obviating aboveground treatment.

The experiments were performed as a series of stimulus-response tests. The stimulus was the injection of the compounds of interest and the response was their concentration history at monitoring locations. The tests included bromide tracer tests to study advection and dispersion; transport tests with the CAHs to study the retardation process due to sorption and to evaluate whether transformation occurred in the absence of active biostimulation; and biostimulation and biotransformation tests to evaluate CAH transformation following the introduction of the primary substrates and electron acceptors.

5.4.1. Results with Methanotrophs

Detailed discussions of the experimental methodology and results of the methanotrophic studies are presented by Roberts et al. (1990) and Semprini et al. (1990, (1991a). Indigenous methanotrophs were stimulated in three successive field seasons through the addition of ground water saturated with methane (16 to 20 mg/l) and oxygen (33 to 38 mg/l), which were introduced into the test zone in alternating pulses without any other supplementary nutrients (N and P).

Figure 5.3 shows the concentration history of methane and oxygen at the S2 observation well during the initial biostimulation experiment along with model simulations of Semprini and McCarty (1991). During the period of 200 to 430 hr, methane and oxygen concentrations rapidly decreased, indicating the growth of methane-utilizers. In order to control the clogging of the injection well and borehole interface, the alternate pulse injection of methane and oxygen containing ground water was initiated at 430 hr, with a pulse cycle time of 4 and 8 hr, respectively. The model simulations, represented by the solid line, matched the field observations using a reasonable set of biological and transport input parameters. Simulation modeling supported the conclusions that methanotrophic bacteria were stimulated in the test zone, that biofouling of the near well-bore region was limited by the pulsing methodology, and that these processes can be simulated when appropriate rate and transport equations are used.

Figure 5.4 shows the response at the S2 well of the target contaminant compounds in the third season and model simulations (Semprini and McCarty, 1992). Transformation of the organic target compounds ensued immediately following the introduction of methane at time zero, increasing with time as the bacterial population grew. Rapid transformation of VC and *trans*-DCE were observed, followed by *cis*-DCE and TCE (not shown). TCA was a ground-water contaminant at the field site, and its possible transformation was followed as well. No transformation of TCA was found; its concentration was about 100 µg/l. Competitive inhibition of VC and *trans*-DCE transformation by methane was indicated in response to the dynamic pulsing of methane and oxygen that was initiated at 20 hr. In order to effectively simulate the field observations, both competitive inhibition kinetics and rate limited sorption were required in the biotransformation model, thus reinforcing the conclusion that these processes were occurring and need to be considered with in-situ remediation. The simulations indicated that the overall rate of decrease in VC concentration may have been limited by the rate of its desorption from the aquifer solids. The simulations also indicated that physical processes, such as desorption, can limit times of cleanup by an enhanced microbial process.

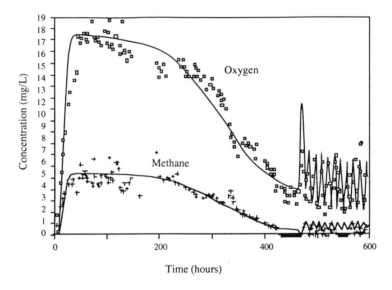

Figure 5.3. Methane and oxygen utilization by methanotrophs at the Moffett test facility (after Semprini and McCarty, 1991).

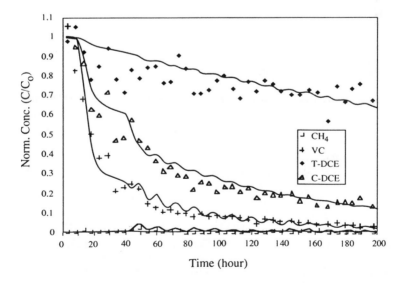

Figure 5.4. CAH transformation by methanotrophs at the Moffett test facility (after Semprini and McCarty, 1992).

The comparison of the cometabolic rate parameters obtained from the modeling exercise indicated that VC and *trans*-DCE were transformed at rates similar to that of methane, the primary substrate added for growth, while *cis*-DCE and TCE were transformed at rates one to two orders of magnitude slower that methane. The simulations indicate that *trans*-DCE concentration decreased more slowly than VC since it was more strongly sorbed (higher K_d), and thus a greater contaminant mass had to be degraded. The order of magnitude difference in rates for *cis*-DCE and *trans*-DCE shows that a small change in chemical structure can have a large effect on the cometabolic transformation rate.

The specific conclusions from this study were:

1) The stimulation of indigenous methanotrophs could be accomplished through methane and DO addition.

2) The rates and extents of transformation of CAHs were compound specific.

3) The percentage transformations achieved in a 2 m biostimulated zone were: TCE, 20%; *cis*-DCE, 50%; *trans*-DCE, 90%; and VC, 95%.

4) The cometabolic transformation was strongly tied to methane utilization; upon stopping methane addition, transformation rapidly ceased.

5) The cometabolic transformation was competitively inhibited by methane, an effect that reduces the transformation rate.

6) Only a temporary enhancement of the cometabolic transformation could be achieved by substituting formate (a noncompetitive substrate) for methane.

7) The rate of transformation was limited by the rate of desorption from the aquifer solids, especially for the more rapidly degraded VC and *trans*-DCE.

8) The results agreed with those obtained in soil microcosm studies, performed under conditions that mimicked the field tests.

5.4.2. Results with Phenol Utilizers

In the methanotrophic studies, limited degradation of TCE and *cis*-DCE was achieved. The objective of this study was to evaluate TO for in-situ biodegradation of TCE, *cis*-DCE and *trans*-DCE at the Moffett test site. This was accomplished through the introduction of phenol as a primary growth substrate and oxygen as an electron acceptor. The evaluation was performed at the same site as the methanotrophic study, using the same experimental methodology and at a similar contaminant concentration range, permitting a direct comparison of the TO system with the MMO system. Active biostimulation was initiated through the pulsing of phenol at time-averaged concentrations ranging from 6 to 12 mg/l.

The concentration responses of DO, TCE, and *cis*-DCE at the SSE2 well, 2 m from the injection well, are shown in Figure 5.5 (Hopkins et al., 1992). The biostimulation with phenol is indicated by the DO decreases, which were small during the periods of low phenol addition but increased after higher phenol concentrations were added. Decreases in *cis*-DCE and TCE concentrations were associated with decreases in DO, indicating cometabolic transformations resulted from biostimulation. Significant degradation of *cis*-DCE, and TCE were observed, with *cis*-DCE being more rapidly degraded than TCE. The *cis*-DCE concentration decreased by approximately 60 to 70% and TCE by 20 to 30% during the period of low phenol addition. Doubling the phenol injection concentration resulted in a greater transformation of both TCE and *cis*-DCE, with 85 to 90%, and over 90%, transformed, respectively. Here, *trans*-DCE (not shown) was the least transformed of the three compounds studied. Upon decreasing the amount of phenol added, the TCE concentration increased, indicating the extent of transformation was related to the amount of phenol added. As with the methanotrophic study, no significant transformation of the TCA ground-water contaminant was observed.

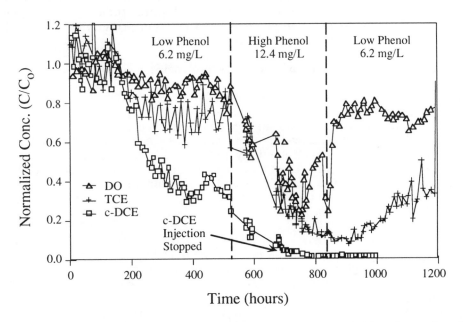

Figure 5.5. **CAH transformation and dissolved oxygen changes resulting from phenol addition at the Moffett test facility (after Hopkins et al., 1992).**

The specific conclusions from this study were:

1) The stimulation of indigenous phenol-utilizers was accomplished through phenol and oxygen addition.

2) The enhanced population effectively degraded TCE and *cis*-DCE but was less effective in degrading *trans*-DCE.

3) Transformations achieved in a 2 m biostimulated zone were: TCE, 85%; and *cis*-DCE over 90%.

4) The cometabolic transformation was competitively inhibited by phenol.

5) The cometabolic transformation was strongly tied to the amount of phenol utilized.

6) The results agreed with microcosm studies that were performed under similar conditions to the field study.

5.4.3. Comparison Between the Methane and Phenol Studies

The pilot-scale studies both demonstrated in-situ biodegradation of CAHs could be achieved. The degradation was shown to be compound specific in both cases. The phenol-utilizers more effectively degraded TCE and *cis*-DCE, while the methane-utilizers more effectively degraded *trans*-DCE, on a percentage basis. Both studies showed that cometabolic transformation was closely associated with primary substrate utilization, and that the primary substrate also competitively inhibited the transformation, slowing the transformation rates. The concentrations of the CAHs were relatively low ($\leq$100 μg/l), thus the results should not be extrapolated to higher concentrations. There is a need for studies over a range of contaminant concentrations. Future studies need to explore CAH concentration effects on degradation efficiency and on other primary substrates that may stimulate more effective oxygenase systems.

5.4.4. Anaerobic Transformation of Carbon Tetrachloride

A Moffett study was also performed to evaluate CT transformation under anaerobic conditions (Semprini et al., 1991b). CT is not transformed through aerobic cometabolism, but anaerobic transformation has been reported and extensively studied. The goals of this field evaluation were: to determine whether reductive transformation of CT could be accomplished under the mildly reducing conditions of denitrification, what factors affect the rates and extents of transformation, and what transformation intermediates might be formed. In addition, other contaminants were present at the site as ground-water contaminants, including TCA and two chlorofluorocarbons, CFC-11, and CFC-113. The disappearance of these compounds was followed as well.

Acetate was first introduced at a time-averaged concentration of 25-46 mg/l, in the presence of the nitrate (25 mg/l) and sulfate (700 mg/l), which were present in the ground water and served as potential electron acceptors. Nitrate utilization commenced immediately after the introduction of acetate and was complete within 100 hours, while acetate utilization also commenced immediately but was expressed somewhat more slowly and less completely because of the stoichiometric excess applied. The onset of CT transformation was observed after approximately 350 hours (Figure 5.6), after which the CT concentrations at the monitoring wells gradually declined, more rapidly at the more distant (S2) well than at the nearer (S1) well. Chloroform (CF) appeared as an intermediate product of the CT transformation at all of the sampling points in an amount corresponding to approximately one-half to two-thirds of the CT that disappeared. The CF response and simulation modeling indicated that CF also was transformed, but more slowly than CT.

The pattern of CT concentrations suggested that the CT transformation proceeded more rapidly further downgradient from the injection well at a location just beyond where nitrate became depleted. To test the hypothesis that the absence of nitrate would enhance the CT transformation, nitrate was removed from the recycled water prior to injection, beginning at 1260 hours. The CT concentration then declined abruptly at the S1 monitoring well. During this period without nitrate feed, the fractional yield of the CF by product declined to about one-third of the CT transformed. Substantial acetate utilization persisted in the absence of nitrate feed, suggesting that sulfate (present at 700 mg/l in the native ground water) may have served as an electron acceptor; however, no sulfide was detected.

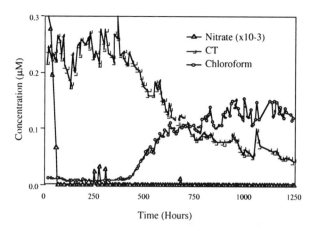

Figure 5.6. CT transformation under anaerobic conditions at the Moffett test site (after Semprini et al., 1991b).

The background contaminants, including 1,1,1-TCA, CFC-11, and CFC-113, were also partially transformed under the influence of anoxic biostimulation.

The specific conclusions from this study were:

1) The stimulation of indigenous acetate-utilizing denitrifiers was accomplished through acetate addition to ground water containing nitrate but no oxygen.

2) Enhanced in-situ reductive transformations can be promoted through the addition of an appropriate primary substrate.

3) Although steady-state conditions were not reached, average transformations achieved over the 2.2 m test distance were: CT, 95%; CFC-11, 68%; CFC-113, 20%, and TCA, 15%.

4) CF was formed as an intermediate product of CT transformation, consistent with laboratory studies.

5.5. PROCEDURES FOR INTRODUCING CHEMICALS INTO GROUND WATER

Perhaps the most significant challenge for in-situ bioremediation of CAHs is the introduction into the subsurface environment of chemicals needed by microorganisms for growth and mixing them with the contaminants to be degraded. Unless a suitable primary substrate for stimulating cometabolic biotransformation of CAHs is already present in the aquifer, one will need to be added. This may not be so difficult in the case of cometabolism under anaerobic conditions, but it may be under aerobic conditions where oxygen must be introduced as well. If methanotrophic cometabolism is desired, then both methane and oxygen must be added. These gases are of limited solubility in water. Therefore, added concentrations, together with other gaseous components, such as molecular nitrogen, collectively must be below the saturation partial pressure in the aquifer, which may not be much higher than 1 atm in shallow ground waters. A common procedure is to mix the chemicals of interest with water and introduce them into the ground water. The introduced water will push away the native ground water containing the contaminants, so that the required mixing between the contaminants and the introduced chemicals will not occur. However, if the contaminants sorb somewhat to the soils, then a benefit will be obtained as the sorbed contaminants will desorb into the introduced water, thus bringing the contaminants and introduced chemicals together.

Another engineering consideration is the problem of excessive microbial growth near the point where the chemicals are introduced into the ground water. Microorganisms will tend to grow near the point of chemical introduction where concentrations are the highest. In order to avoid such clogging, a strategy is needed that will make growth difficult near the point of injection. The periodic introduction of inhibitory chemicals such as chlorine or ozone may be required. A strategy such as pulsing of the primary substrate and oxygen, as performed in the Moffett Field experiments described previously, so that they do not occur together near the injection point is a possibility. Full-scale experience with such strategies is limited and so documented guidelines for application are not available.

As with other forms of bioremediation presented in this section, delivery of essential nutrients is a prerequisite for effective in-situ bioremediation. One potential method for accomplishing the mixing of nutrients with ground-water contaminants is to use a pump-and-treat extraction and injection system with bioremediation added, as depicted in Figure 5.7 (McCarty et al., 1991). Here, all of the biological treatment is carried out in the aquifer itself. Ground water is extracted through a series of wells spanning the contaminant plume in a direction perpendicular to that of ground-water flow. At the surface, methane and oxygen are added to the extracted ground water, either together or in alternating pulses, depending upon the extent to which a stimulated methanotrophic population has been developed.

The alternating pulses are used to distribute the microbial growth throughout the test zone. The ground water containing the appropriate primary substrates and electron acceptors (i.e., methane and oxygen) and extracted contaminants are reinjected into the treatment zone through a series of wells distributed parallel to the extraction wells. In the subsurface biotreatment zone, both the in-place contaminants and the reinjected contaminants are biologically degraded. Another alternative is to use a combination of aboveground treatment and in-situ treatment. In this case, the contaminants would be removed at the surface, and only the required amendments would be added to the extracted ground water prior to reinjection.

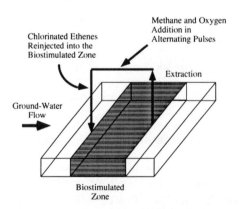

Figure 5.7.   **Pump-extract-reinject method for mixing of chemicals with ground water (after McCarty et al., 1991).**

The above in-situ biotreatment system was directly compared with a pump-and-treat system through model simulations. In the latter case, an unspecified form of surface treatment was assumed to remove contaminants quantitatively before the water is reinjected. Otherwise, the two systems were assumed to operate identically, allowing direct comparison through model simulations (Semprini and McCarty, 1991, 1992). The results of model simulations are shown in Figure 5.8 for VC, a weakly sorbed contaminant that is rapidly degraded by the methanotrophic process. Here the in-situ process was shown to be as effective as pump and treat. The model simulations of Figure 5.8 also indicate the advisability of recycling the contaminants through the treatment zone rather than removing them through treatment at the surface. The line with triangles (Biostim+pump) shows a simulation in which the contaminants in the extracted water were removed using surface treatment before reinjection. Methane and oxygen were added to the surface-treated reinjected water. Some, but not a significant, enhancement of removal was achieved by adding surface treatment. Moreover, the in-situ bioremediation process also degraded the reinjected contaminants to nontoxic end products, which is an advantage over some forms of surface treatment.

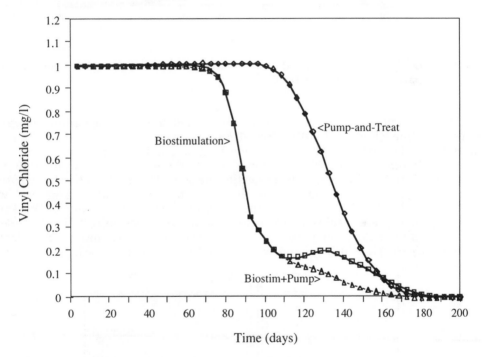

Figure 5.8. Simulation modeling of pump-extract-reinject method for methanotrophic cometabolism of VC (after McCarty et al., 1991).

Model simulations are quite useful in evaluating the different approaches and potential effectiveness of proposed treatment schemes. Simulations for *trans*-DCE (Figure 5.9), a more strongly sorbed compound than VC but one that is as rapidly degraded by methanotrophs, shows the in-situ process becoming even more attractive than pump and treat. However, for compounds that were less effectively degraded by the methanotrophic process, such as *cis*-DCE and TCE, in-situ treatment was found through simulation modeling to be less effective in reducing the cleanup times or the amount of water extracted compared to normal pump-and-treat procedures.

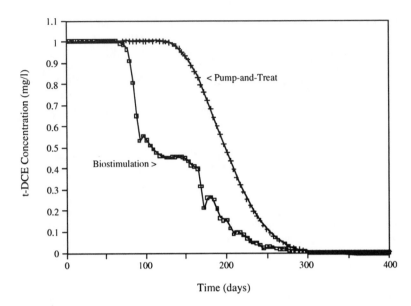

Figure 5.9. Simulation modeling of pump-extract-reinject method for methanotrophic cometabolism of *trans*-DCE.

Another possible system for delivering the needed chemicals is subsurface ground-water recirculation (Figure 5.10). This eliminates the need to pump contaminated ground water to the surface. This mixing method, which is under development, uses a subsurface recirculation unit with an upper and lower screen, and a pump to induce flow through the unit and through the porous formation. For methanotrophic treatment, methane and oxygen would be introduced directly into the recirculating ground water. This method would eliminate pumping the contaminated ground water to the surface, surface treatment, and subsequent reinjection. One possibility is that several recirculation units could span perpendicularly across a plume, making a biologically reactive barrier through which ground water would flow and be treated.

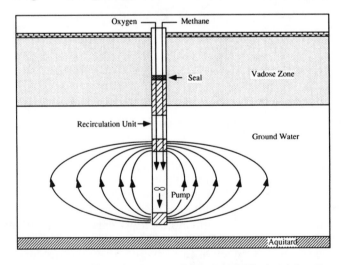

Figure 5.10. Subsurface recirculation system for chemical introduction and mixing with ground-water contaminants.

The mixing system illustrated in Figure 5.7 might be used not only with other forms of aerobic treatment, but also with anaerobic treatment as well. One possibility is to use in-situ treatment to augment a pump-and-treat remediation that is already in place.

5.6.　THE EFFECT OF SITE CONDITIONS ON REMEDIATION POTENTIAL

Figure 5.11 illustrates contamination of soil and ground water by leakage from a storage tank, a common way by which contamination with liquids occurs (McCarty, 1990). As the liquid is pulled downward by gravity, residuals left behind contaminate the surface soil, the unsaturated (vadose) zone, and finally the aquifer containing the ground water itself. After the leakage is found and stopped, and the most highly contaminated soil around the tank is excavated, one must then deal with a lower-concentration residual in the soil, the vadose zone, and the ground water. If the contaminating liquid is a mixture of many different compounds, then each may move and be transformed at different rates. Biological and chemical transformations may not lead to mineralization, but may result in producing other organic chemicals that may be either less or more harmful than the original. Organics may become strongly sorbed onto subsurface minerals or may penetrate into cracks so that they are not accessible by microorganisms or their enzymes.

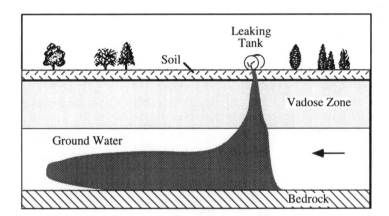

Figure 5.11. CAH contamination in a relatively homogeneous subsurface environment (after McCarty, 1990).

The relatively homogeneous subsurface environment indicated in Figure 5.11 is ideal, but seldom encountered. In such a case, ground-water flow direction and rate might be determined from relatively few observations of piezometric heads and data from pumping tests. Subsurface environments often are much more complex than this, some perhaps as illustrated in Figure 5.12. Layering of permeable (sands and gravels) and less-permeable (silts, clays, rock) strata is common and may contain discontinuities that could result from faults or large-scale stratigraphic features. Conductivity of water and contaminants through rocks and other such barriers may result from joints and fractures that are difficult to locate and to describe. The mixture of gravel, sand, silt, clay, and organic matter of which the subsurface environment consists can vary widely from location to location, as can the grain-size distribution and mineral composition within each broad class of subsurface strata.

In addition, abandoned wells can often provide passageways between separated aquifers. Recalcitrance of contaminants in such systems may result from high concentrations that are toxic, from presence of the organics in fissures, strong sorption to particle surfaces, or diffusion into small pore spaces in minerals, rendering the contaminants inaccessible to microorganisms and their enzymes. Sorbed compounds often desorb slowly, and this often becomes the limiting factor affecting rates of biodegradation as well as removal by pump-and-treat methods (McCarty, 1990). Recalcitrance may also result from insufficient levels of required nutrients, such as nitrogen and phosphorus for bacterial growth (Knox et al., 1985; Thomas and Ward, 1989). For aerobic treatment, optimal concentrations of ammonia or nitrate nitrogen are in the range of 2 to 8 pounds per 100 pounds of organic material, while inorganic phosphorus requirements are about one-fifth of this (McCarty, 1988). When these nutrients are below optimum levels, rates of biodegradation slow considerably and may be more dependent upon rates of nitrogen and phosphorus regeneration than on other factors. With anaerobic degradation, nutrient needs are generally less, but organism growth rates are slower. The absence of suitable electron acceptors is another factor that can affect biodegradability. For aromatic hydrocarbons, degradation rates are generally enhanced through aerobic decomposition. Thus, introduction of oxygen can be useful. Generally, the quantity of oxygen required is similar to the mass of contaminants present. In complex subsurface systems, getting the oxygen to the areas of need can prove difficult.

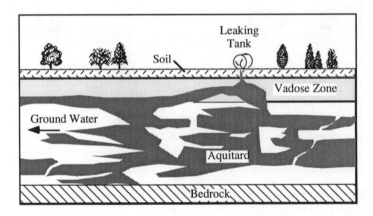

Figure 5.12. CAH contamination in a relatively nonhomogeneous subsurface environment (after McCarty, 1990).

Frequently, when environmental conditions are not appropriate for biodegradation to occur, potential solutions often involve addition of chemicals (Knox et al., 1985; Roberts et al., 1989; Thomas and Ward, 1989; McCarty et al., 1991). This is perhaps not difficult with surface contamination but may be nearly impossible with some subsurface contamination, depending upon the hydrogeology. With the latter, conditions that make pump and treat difficult render efforts at bioremediation difficult as well. If it is difficult to pump contaminants out of the ground, then it is also difficult to pump chemicals or microorganisms into the ground to reach the contaminants. In such cases, biological approaches may not offer significant time advantages over pump and treat. The main advantage of bioremediation is likely to be an environmental one, i.e., the contaminants are destroyed with a minimum of disruption of the surface environment. In some cases, costs may be significantly reduced as well.

In some cases, proper environmental conditions may be obtained by moving contaminants to effect dilution or mixing with natural chemicals in the subsurface system. Dilution by mixing of contaminated and uncontaminated ground water can reduce contaminant toxicity. Also with dilution, alternate electron acceptors such as oxygen or nitrates, or essential nutrients such as nitrates, phosphates and iron, that are present in uncontaminated water, may be brought together with the contaminants for better biodegradation. Again, better methodologies for predicting the outcome of this strategy are needed.

Where environmental conditions are suitable and where the proper microbial populations are present, complete mineralization of organic contaminants can occur, even within the most complex hydrogeological environments. Even where environmental conditions are not ideal, degradation of many organic chemicals may take place at reduced rates, with half-lives on the order of one or two years. In such cases, the correct strategy may be to leave the contaminants alone and allow the problem to be rectified by natural processes. Environmentally, this may be the best position to take. The difficulty here is in obtaining evidence that would convince us, the regulatory authorities, and the public that such natural processes are indeed occurring. Also difficult is making good estimates of the time-frame for natural purification to occur. Currently, we do not know what evidence to collect to prove the occurrence of natural degradative processes, nor how to collect it. This is a most important area of need.

5.6.1. Microorganism Presence

With contaminants that are known to be readily biodegradable, the absence of a suitable microbial population may also be a factor. Methodologies for determining microorganism presence are under development. Some include the simple exposure of aseptically obtained soil to the contaminants of concern under ideal chemical conditions for biodegradation. If the microorganisms are naturally present, then degradation of the contaminant will occur. Other approaches are to attempt to identify the presence of species known to biodegrade the compounds of interest, or to use molecular probes that can identify the presence of specific microorganisms, nucleic acid sequences, or enzymes that are key to compound degradation. These more sophisticated techniques are not yet fully developed, but may offer promise for the future.

If appropriate organisms are not present, then they may be introduced into the surface or subsurface environment (Omenn et al., 1988). Such organisms may be natural, but not ubiquitous in nature. Their growth and introduction into a new system may thus be acceptable. An important question is whether such specialized organisms can survive in the new environment, and if so, can they be transported to the place of need? If the hydrogeology is complex, then this may be most difficult. In other research, attempts are being made to engineer microorganisms that are capable of degrading organic compounds that are inherently recalcitrant. The potential use of such organisms raises societal concerns as well as the physical and biological barriers to successful organism introduction into the environment. Nevertheless, such approaches deserve to be explored as they will add to our overall knowledge of the biodegradation process.

5.7.   SUMMARY

In-situ biodegradation of most CAHs depends upon cometabolism and can be carried out under aerobic or anaerobic conditions. Cometabolism requires that an appropriate primary substrate be added to the aquifer, and perhaps an electron acceptor such as oxygen or nitrate for its oxidation. Only a few CAHs can serve as primary substrates for biodegradation. In order to apply in-situ bioremediation, conditions must be appropriate. The aquifer should be relatively homogeneous so that chemicals can be mixed with ground water. Sufficient primary substrate must be added to satisfy the needs of the respective bacteria. Generally, the more highly chlorinated CAHs should be converted by anaerobic in-situ treatment to less chlorinated forms that can be degraded through aerobic cometabolism. It is necessary to determine whether the appropriate microorganisms are present as indigenous organisms in the aquifer. Generally, this requires laboratory studies on aseptically obtained aquifer material.   Sufficient characterization of the aquifer is desirable so that the injection and distribution of chemicals can be modeled and decomposition of CAHs can be reasonably well-predicted.

The formation of halogenated intermediate products, which may be of public health concern, poses an obstacle to the deployment of the anaerobic approach for aquifer bioremediation with the more highly chlorinated species. Recent laboratory and field investigations, however, show PCE and TCE can be completely dehalogenated to ethene, which is encouraging. Research work therefore needs to focus on determining how effectively enhanced reductive dehalogenation to nontoxic end products can be accomplished in the complex subsurface environment and the potential benefits as well as disadvantages of forming less halogenated intermediates.

Until full-scale experience is available, the best approach might be to attempt remediating sites that are relatively simple hydrogeologically and contain more readily degradable contaminants.   An ideal case would be to degrade VC through aerobic cometabolism with methanotrophs. VC is difficult to remove with the normal pump-and-treat system because it is a known human carcinogen with relatively low MCL, and it does not sorb well to activated carbon or other sorbers. Thus, surface treatment is difficult and expensive. However, VC can be used as a primary substrate, if the appropriate organisms are present, or at least can be destroyed through cometabolism by methanotrophic bacteria. Here, the ratio of methane addition and VC degradation is quite low, about two kg of methane are required per kg of VC destroyed. Experience is also needed with systems for introducing chemicals into the subsurface environment and for mixing them with the contaminants of concern. Once experience with the easier cases is available, then application in more complex situations can be attempted. Without full-scale application, little can be said about the cost of such treatment. Thus, there is much yet to be learned.

ACKNOWLEDGEMENT

This report contains information that resulted from studies conducted through the Western Region Hazardous Substance Research Center through EPA Grant No. R815738.

REFERENCES

Alvarez-Cohen, L.M., and P.L. McCarty. 1991a. Effects of toxicity, aeration, and reductant supply on trichloroethylene transformation by a mixed methanotrophic culture. *Appl. Environ. Microbiol.* 57(1):228-235.

Alvarez-Cohen, L.M., and P.L. McCarty. 1991b. Product toxicity and cometabolic competitive inhibition modeling of chloroform and trichloroethylene transformation by methanotrophic resting cells. *Appl. Environ. Microbiol.* 57:(4)1031-1037.

Arciero, D., T. Vannelli, M. Logan, and A.B. Hooper. 1989. Degradation of trichloroethylene by the ammonia-oxidizing bacterium *Nitrosomonas europaea*. *Biochem. Biophys. Res. Commun*. 159:640-643.

Bouwer, E.J., and J.P. Wright. 1988. Transformations of trace halogenated aliphatics in anoxic biofilm columns. *J. Contaminant Hydrol*. 2:155-169.

Bouwer, E.J., B.E. Rittmann, and P.L. McCarty. 1981. Anaerobic degradation of halogenated 1- and 2-carbon organic compounds. *Environ. Sci. Technol*. 15(5):596-599.

Brunner, W., and T. Leisinger. 1978. Bacterial degradation of dichloromethane. *Experientia*. 34:1671.

Brunner, W., D. Staub, and T. Leisinger. 1980. Bacterial degradation of dichloromethane. *Appl. Environ. Microbiol.* 40(5):950-958.

Cline, P.V., and J.J. Delfino. 1989. Transformation kinetics of 1,1,1-trichloroethane to the stable product 1,1-dichloroethene. In: *Biohazards of Drinking Water Treatment*. Lewis Publishers, Inc. Chelsea, Michigan. pp. 47-56.

Curtis, G.P., P.V. Roberts, and M. Reinhard. 1986a. A natural gradient experiment on transport in a sand aquifer. IV. Sorption of organic solutes and its relation to mobility. *Water Resour. Res*. 22(13):2059-2067.

Curtis, G.P., M. Reinhard, and P.V. Roberts. 1986b. Sorption of hydrophobic organic compounds by sediments. In: *Geochemical Processes at Mineral Surfaces*. Eds., J.A. Davis and K.F. Hayes. ACS Symposium Series No. 323. American Chemical Society. Washington, DC. pp. 191-216.

DiStefano, T.D., J.M. Gossett, and S.H. Zinder. 1991. Reductive dechlorination of high concentrations of tetrachloroethene to ethene by an anaerobic enrichment culture in the absence of methanogenesis. *Appl. Environ. Microbiol.* 57(8):2287-2292.

Ewers, J., W. Clemens, and H.J. Knackmuss. 1991. Biodegradation of chloroethenes using isoprene as co-substrate. In: *Proceedings of International Symposium: Environmental Biotechnology*. European Federation of Biotechnology. Oostende, Belgium. April 22-25. pp. 77-83.

Fogel, M.M., A.R. Taddeo, and S. Fogel. 1986. Biodegradation of chlorinated ethenes by a methane-utilizing mixed culture. *Appl. Environ. Microbiol.* 51(4):720-724.

Folsom, B.R., P.J. Chapman, and P.H. Pritchard. 1990. Phenol and trichloroethylene degradation by *Pseudomonas cepacia* G4: Kinetics and interactions between substrates. *Appl. Environ. Microbiol*. 56(5):1279-1285.

Fox, B.G., J.G. Borneman, L.P. Wackett, and J.D. Lipscomb. 1990. Haloalkene oxidation by the soluble methane monooxygenase from *Methylosinus trichosporium* OB3b: Mechanistic and environmental implications. *Biochemistry*. 29:6419-6427.

Freedman, D.L., and J.M. Gossett. 1989. Biological reductive dechlorination of tetrachloroethylene and trichloroethylene to ethylene under methanogenic conditions. *Appl. Environ. Microbiol*. 55(9):2144-2151.

Freedman, D.L., and J.M. Gossett. 1991. Biodegradation of dichloromethane in a fixed film reactor under methanogenic conditions. In: *On-Site Bioremediation Processes for Xenobiotic and Hydrocarbon Treatment*. Eds., R.E. Hinchee and R.F. Olfenbuttel. Butterworth-Heinemann. Boston, Massachusetts. pp. 113-133.

Freeze, R.A., and J.A. Cherry. 1979. *Groundwater*. Prentice-Hall, Inc. Englewood Cliffs, New Jersey.

Goltz, M.N., and P.V. Roberts. 1986. Interpreting organic solute transport data from a field experiment using physical nonequilibrium models. *J. Contaminant Hydrol*. 1:77-93.

Goltz, M.N., and P.V. Roberts. 1987. Using the method of moments to three-dimensional, diffusion-limited transport from temporal and spatial perspectives. *Water Resour. Res*. 23(8):1575-1585.

Harker, A.R., and Y. Kim. 1990. Trichloroethylene degradation by two independent aromatic-degrading pathways in *Alcaligenes eutrophus* JMP134. *Appl. Environ. Microbiol*. 56(4):1179-1181.

Hartmans, S., J.A.M. de Bont, J. Tramper, and K.Ch.A.M. Luyben. 1985. Bacterial degradation of vinyl chloride. *Biotechnol. Letters*. 7(6):383-388.

Hartmans, S., and J.A.M. de Bont. 1992. Aerobic vinyl chloride metabolism in *Mycobacterium aurum* L1. *Appl. Environ. Microbiol*. 58(4):1220-1226.

Henry, S.M. 1991. *Transformation of Trichloroethylene by Methanotrophs from a Groundwater Aquifer*. Ph.D. Thesis. Stanford University. Stanford, California.

Henry, S.M., and D. Grbic-Galic. 1990. Effect of mineral media on trichloroethylene oxidation by aquifer methanotrophs. *Microb. Ecol*. 20:151-169.

Henry, S.M., and D. Grbic-Galic. 1991a. Influence of endogenous and exogenous electron donors and trichloroethylene oxidation toxicity on trichloroethylene oxidation by methanotrophic cultures from a groundwater aquifer. *Appl. Environ. Microbiol*. 57(1):236-244.

Henry, S.M., and D. Grbic-Galic. 1991b. Inhibition of trichloroethylene oxidation by the transformation intermediate carbon monoxide. *Appl. Environ. Microbiol*. 57(6):1770-1776.

Henson, J.M., M.V. Yates, and J.W. Cochran. 1989. Metabolism of chlorinated methanes, ethanes, and ethylenes by a mixed bacterial culture growing on methane. *J. Ind. Microbiol.* 4:29-35.

Hopkins, G.D., L. Semprini, and P.L. McCarty. 1992. Evaluation of enhanced in situ aerobic biodegradation of trichloroethylene, and *cis*-and-*trans*-1,2-dichloroethylene by phenol utilizing bacteria. Abstract: *Symposium on Bioremediation of Hazardous Wastes, EPA's Biosystems Technology Development Program.* Chicago, Illinois. May 5-6.

Janssen, D.B., A. Scheper, L. Dijkhuizen, and B. Witholt. 1985. Degradation of halogenated aliphatic compounds by *Xanthobacter autotrophicus* GJ10. *Appl. Environ. Microbiol.* 49(3):673-677.

Jeffers, P.M., L.M. Ward, L.M. Woytowitch, and N.L. Wolfe. 1989. Homogeneous hydrolysis rate constants for selected chlorinated methanes, ethanes, ethenes, and propanes. *Environ. Sci. Technol.* 23(8):965-969.

Karickhoff, S.W., D.S. Brown, and T.A. Scott. 1979. Sorption of hydrophobic pollutants on natural sediments. *Water Research.* 13:241-248.

Klecka, G.M. 1982. Fate and effects of methylene chloride in activated sludge. *Appl. Environ. Microbiol.* 44(3):701-707.

Knox, R.C., L.W. Canter, D.F. Kincannon, E.L. Stover, and C.H. Ward. 1985. *State-of-the-Art of Aquifer Restoration.* EPA/600/S2-84/182. U.S. Environmental Protection Agency. Cincinnati, Ohio.

Kohler-Staub, D., and T. Leisinger. 1985. Dichloromethane dehalogenase of *Hyphomicrobium* sp. Strain DM2. *J. Bacteriol.* 162:676-681.

Lanzarone, N.A., and P.L. McCarty. 1990. Column studies on methanotrophic degradation of trichloroethene and 1,2-dichloroethane. *Ground Water.* 28(6):910-919.

LaPat-Polasko, L.T., P.L. McCarty, and A.J.B. Zehnder. 1984. Secondary substrate utilization of methylene chloride by an isolated strain of *Pseudomonas* sp. *Appl. Environ. Microbiol.* 47(4):825-830.

Little, C.D., A.V. Palumbo, S.E. Herbes, M.E. Lidstrom, R.L. Tyndall, and P.J. Gilmer. 1988. Trichloroethylene biodegradation by a methane-oxidizing bacterium. *Appl. Environ. Microbiol.* 54(4):951-956.

Mackay, D.M., P.V. Roberts, and J.A. Cherry. 1985. Transport of organic contaminants in groundwater: A critical review. *Environ. Sci. Technol.* 19(5):384-392.

Mackay, D.M., D.L. Freyberg, P.V. Roberts, and J.A. Cherry. 1986. A natural-gradient experiment on solute transport in a sand aquifer. I. Approach and overview of plume movement. *Water Resour. Res.* 22(13):2017-2029.

Mayer, K.P., D. Grbic-Galic, L. Semprini, and P.L. McCarty. 1988. Degradation of trichloroethylene by methanotrophic bacteria in a laboratory column of saturated aquifer material. *Wat. Sci. Tech.* (Great Britain) 20(11/12):175-178.

McCarty, P.L.   1988.   Bioengineering issues related to in-situ remediation of contaminated soils and groundwater. In: *Environmental Biotechnology*. Ed., G.S. Omenn. Plenum Publishing Corp. New York, New York. pp. 143-162.

McCarty, P.L. 1990.   Scientific limits to remediation of contaminated soils and groundwater. In: *Ground Water and Soil Contamination Remediation: Toward Compatible Science, Policy, and Public Perception*. Washington, DC. National Academy Press. pp. 38-52.

McCarty, P.L., M. Reinhard, and B.E. Rittmann. 1981. Trace organics in groundwater. *Environ. Sci. Technol.* 15(1):40-51.

McCarty, P.L., L. Semprini, M.E. Dolan, T.C. Harmon, C. Tiedeman, and S.M. Gorelick. 1991. In-situ methanotrophic bioremediation of contaminated groundwater at St. Joseph, Michigan. In: *On-Site Bioremediation Processes for Xenobiotic and Hydrocarbon Treatment*. Eds., R.E. Hinchee and R.F. Olfenbuttel. Butterworth-Heinemann. Boston, Massachusetts. pp. 16-40.

McCarty, P.L., P.V. Roberts, M. Reinhard, and G. Hopkins. 1992. Movement and transformations of halogenated aliphatic compounds in natural systems. In: *Fate of Pesticides and Chemicals in the Environment*. Ed., J.L. Schnoor. John Wiley & Sons, Inc. New York, New York. pp. 191-209.

Miller, R.E., and F.P. Guengerich.   1982.   Oxidation of trichloroethylene by liver microsomal cytochrome P-450: Evidence for chlorine migration in a transition state not involving trichloroethylene oxide. *Biochemistry*. 21:1090-1097.

National Academy of Sciences (NAS). 1984. *Groundwater Contamination*. National Academy Press. Washington, DC.

Nelson, M.J.K., S.O. Montgomery, E.J. O'Neill, and P.H. Pritchard. 1986. Aerobic metabolism of trichloroethylene by a bacterial isolate. *Appl. Environ. Microbiol.* 52(2):383-384.

Nelson, M.J.K., S.O. Montgomery, W.R. Mahaffey, and P.H. Pritchard.   1987. Biodegradation of trichloroethylene and involvement of an aromatic biodegradative pathway. *Appl. Environ. Microbiol.* 53(5):949-954.

Nelson, M.J.K., S.O. Montgomery, and P.H. Pritchard. 1988. Trichloroethylene metabolism by microorganisms that degrade aromatic compounds. *Appl. Environ. Microbiol.* 54(2):604-606.

Oldenhuis, R., R.L.J.M. Vink, D.B. Janssen, and B. Witholt. 1989. Degradation of chlorinated aliphatic hydrocarbons by *Methylosinus trichosporium* OB3b expressing soluble methane monooxygenase. *Appl. Environ. Microbiol.* 55(11):2819-2826.

Oldenhuis, R., J.Y. Oedzes, J.J. van der Waarde, and D.B. Janssen. 1991. Kinetics of chlorinated hydrocarbon degradation by *Methylosinus trichosporium* OB3b and toxicity of trichloroethylene. *Appl. Environ. Microbiol.* 57(7):7-14.

Omenn, G.S., R. Colwell, A.M. Chakrabarty, M. Lewis, and P. McCarty (Eds.) 1988. *Environmental Biotechnology, Reducing Risks From Environmental Chemicals Through Biotechnology*. Plenum Press. New York, New York.

Rittmann, B.E., and P.L. McCarty. 1980a. Utilization of dichloromethane by suspended and fixed-film bacteria. *Appl. Environ. Microbiol*. 39(6):1225-1226.

Rittmann, B.E., and P.L. McCarty. 1980b. Model of steady-state biofilm kinetics. *Biotech. Bioengin*. 22:2343-2357.

Roberts, P.V., P.L. McCarty, M. Reinhard, and J. Schreiner. 1980. Organic contaminant behavior during groundwater recharge. *J. Water Poll. Contr. Fed*. 52:134-147.

Roberts, P.V., M. Reinhard, and A.J. Valocchi. 1982. Movement of organic contaminants in groundwater. *J. Am. Water Works Assoc*. 74(8):408-413.

Roberts, P.V., M.N. Goltz, and D.M. Mackay. 1986. A natural-gradient experiment on solute transport in a sand aquifer. III. Retardation estimates and mass balances for organic solutes. *Water Resour. Res*. 22(13):2047-2058.

Roberts, P.V., L. Semprini, G.D. Hopkins, D. Grbic-Galic, P.L. McCarty, and M. Reinhard. 1989. *In Situ Aquifer Restoration of Chlorinated Aliphatics by Methanotrophic Bacteria*. EPA/600/2-89/033. Center for Environmental Research Information. Cincinnati, Ohio.

Roberts, P.V., G.D. Hopkins, D.M. Mackay, and L. Semprini. 1990. A field evaluation of in-situ biodegradation of chlorinated ethenes: Part 1. Methodology and field site characterization. *Ground Water*. 28(4):591-604.

Schwarzenbach, R.P., and J. Westall. 1981. Transport of nonpolar organic compounds from surface water to groundwater laboratory sorption studies. *Environ. Sci. Technol*. 15(11):1360-1367.

Semprini, L., and P.L. McCarty. 1991. Comparison between model simulations and field results for in-situ biorestoration of chlorinated aliphatics: Part 1. Biostimulation of methanotrophic bacteria. *Ground Water*. 29(3):365-374.

Semprini, L., and P.L. McCarty. 1992. Comparison between model simulations and field results of in-situ biorestoration of chlorinated aliphatics: Part 2. Cometabolic transformations. *Ground Water*. 30(1):37-44.

Semprini, L., G.D. Hopkins, P.V. Roberts, D. Grbic-Galic, and P.L. McCarty. 1991a. A field evaluation of in-situ biodegradation of chlorinated ethenes: Part 3. Studies of competitive inhibition. *Ground Water*. 29(2):239-250.

Semprini, L., G.D. Hopkins, D.B. Janssen, M. Lang, P.V. Roberts, and P.L. McCarty. 1991b. *In-situ Biotransformation of Carbon Tetrachloride Under Anoxic Conditions*. EPA/2-90/060. Robert S. Kerr Environmental Research Laboratory. Ada, Oklahoma.

Semprini, L., Grbic-Galic, D., McCarty, P.L. and P.V. Roberts. 1992. *Methodologies for Evaluating In-situ Bioremediation of Chlorinated Solvents*. USEPA 600/R-92/042. Robert S. Kerr Environmental Research Laboratory. Ada, Oklahoma.

Semprini, L., P.V. Roberts, G.D. Hopkins, and P.L. McCarty. 1990. A field evaluation of in-situ biodegradation of chlorinated ethenes: Part 2. Results of biostimulation and biotransformation experiments. *Ground Water*. 28(5):715-727.

Stucki, G., R. Galli, H.R. Ebersold, and T. Leisinger. 1981. Dehalogenation of dichloromethane by cell extracts of *Hyphomicrobium* DM2. *Arch. Microbiol.* 130:366-371.

Stucki, G., U. Krebser, and T. Leisinger. 1983. Bacterial growth on 1,2-dichloroethane. *Experientia*. 39:1271-1273.

Thomas, J., and C. Ward. 1989. In situ biorestoration of organic contaminants in the subsurface. *Environ. Sci. Technol.* 23(7): 760-766.

Tsien, H.-C., G.A. Brusseau, R.S. Hanson, and L.P. Wackett. 1989. Biodegradation of trichloroethylene by *Methylosinus trichosporium* OB3b. *Appl. Environ. Microbiol.* 55(12):3155-3161.

Vannelli, T., M. Logan, D.M. Arciero, and A.B. Hooper. 1990. Degradation of halogenated aliphatic compounds by the ammonia-oxidizing bacterium *Nitrosomonas europaea*. *Appl. Environ. Microbiol.* 56(4):1169-1171.

Vogel, T.M., C.S. Criddle, and P.L. McCarty 1987. Transformations of halogenated aliphatic compounds. *Environ. Sci. Technol.* 21(8):722-736.

Wackett, L.P., and D.T. Gibson. 1988. Degradation of trichloroethylene by toluene dioxygenase in whole-cell studies with *Pseudomonas putida* F1. *Appl. Environ. Microbiol.* 54(7):1703-1708.

Wackett, L.P., G.A. Brusseau, S.R. Householder, and R.S. Hanson. 1989. Survey of microbial oxygenases: Trichloroethylene degradation by propane-oxidizing bacteria. *Appl. Environ. Microbiol.* 55(11):2960-2964.

Wilson, J.T., and B.H. Wilson. 1985. Biotransformation of trichloroethylene in soil. *Appl. Environ. Microbiol.* 49(1):242-243.

SECTION 6

BIOVENTING OF CHLORINATED SOLVENTS FOR GROUND-WATER CLEANUP THROUGH BIOREMEDIATION

John T. Wilson
Don H. Kampbell
U.S. Environmental Protection Agency
Robert S. Kerr Environmental Research Laboratory
P.O. Box 1198
Ada, Oklahoma 74820
Telephone: (405)436-8532, 8564
Fax: (405)436-8529

6.1. FUNDAMENTAL PRINCIPLES

Chlorinated solvents such as tetrachloroethylene, trichloroethylene, carbon tetrachloride, chloroform, 1,2-dichloroethane, and dichloromethane (methylene chloride) can exist in contaminated subsurface material as (1) the neat oil, (2) a component of a mixed oily waste, (3) a solution in soil water, or (4) a vapor in soil air. Spills of such materials to subsurface material are frequently treated by soil vacuum extraction to remove volatile oils and vapors in soil air, or by air stripping of contaminated ground water. Both physical treatment processes produce a waste stream of contaminant vapors in air. At present, these wastes are discharged to the atmosphere, or treated with activated carbon, catalytic combustion or incineration.

Bioventing refers to biological treatment of oils in the vadose zone, supported by oxygen delivered to the contaminated subsurface material through the advective flow of air. In areas contaminated with oily phase material, oxygen delivered in air can support direct biological degradation of the oily material. Any material volatilized into the air may be swept away before treatment can occur (see section 3.2 for a discussion). If there is adequate residence time of the vapors in material without oily phase contaminants, the contaminant vapors can be degraded as the air moves away from the source areas.

In theory, a similar process can be used to support biological destruction of certain chlorinated solvents. Many chlorinated solvents are not subject to direct biodegradation, but can be cooxidized during microbial growth on another hydrocarbon. Frequently, the extent of destruction of the chlorinated solvent is limited by the low solubility of the growth substrate, and of oxygen, in water. Air is an ideal medium to deliver oxygen and hydrocarbons for microbial growth to contaminated subsurface environments.

The chlorinated solvents, particularly trichloroethylene, are toxic to organisms that cooxidize them. Toxic effects of trichloroethylene become important above 6 mg/l water or 2 mg/l air (Table 6.2; Broholm et al., 1990; Broholm et al., 1991). Air or water in contact with oily phase trichloroethylene frequently exceeds the toxic limit. Further, biodegradation supported in one pass through the unsaturated zone may not reduce the concentration of contaminants to acceptable limits. In the near term, successful implementation of bioventing for chlorinated solvents will most likely include (1) physical transfer of the contaminant to air through soil vacuum extraction or air sparging of ground water, (2) dilution of contaminants, if necessary, by addition of make up air, followed by reinjection into subsurface material that is not contaminated with oily phase solvent. This idealized implementation is illustrated in Figure 6.1.

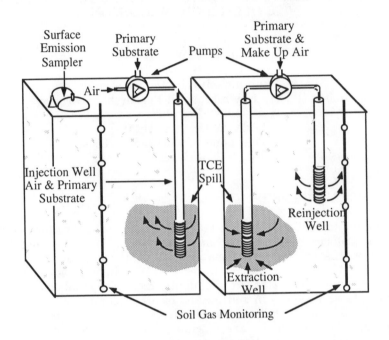

Figure 6.1.   Two hypothetical implementations of in-situ bioventing of chlorinated solvents.

6.2.   MATURITY OF THE TECHNOLOGY

As of this writing (1993) the technology is emerging.   Bench-scale systems have been evaluated in a number of laboratories, but no pilot- or field-scale system is yet in place. The technology is essentially a blend of bioventing as an engineered activity, and biotechnology for cooxidation of chlorinated solvents. Progress in bioventing is rapid, and considerable effort is being expended on the biology, physiology, and biochemistry of chlorinated solvent cooxidation in ground water.   Bioventing of chlorinated solvents awaits the linkage of bioventing and the microbiology of cooxidation of the chlorinated solvents.

There is a good prospect that the States and U.S. Environmental Protection Agency will see permit applications for bioventing of chlorinated solvents starting in 1993 or 1994.

6.3.   PRIMARY REPOSITORIES OF EXPERTISE

The best University research to date has been done at the University of Texas at Austin in the laboratories of Gerald Speitel and Ray Loehr. Don Kampbell at the Kerr Laboratory at Ada, Oklahoma, did the seminal work in this area, and continues an active laboratory-scale program. Harvey Read and Thomas Stocksdale with S.C. Johnson & Son, Inc. (Racine, Wisconsin) and Hinrich L. Bohn of the University of Arizona (Tuscon, Arizona) have the most experience with field-scale biopiles designed to oxidize hydrocarbon vapors. These biopiles are very similar to systems that would be designed for cooxidation of chlorinated solvents.

Hinrich L. Bohn
Dept. of Soil & Water Science
University of Arizona
Tuscon, AZ 85721
Phone: (602)621-1646
Fax: (602)621-1647

Harvey W. Read
S.C. Johnson & Son, Inc.
1525 Howe Street
Racine, WI 53403
Phone: (414)631-2000
Fax: (414)631-2167

Don H. Kampbell
U.S. EPA
Robert S. Kerr Laboratory
Ada, OK 74820
Phone: (405)436-8564
Fax: (405)436-8529

Gerald E. Speitel
Dept. of Civil Engineering
The University of Texas at Austin
Austin, TX 78712-1076
Phone: (512)471-5602
Fax: (512)471-0592

Raymond C. Loehr
Dept. of Civil Engineering
The University of Texas at Austin
Austin, TX 78712-1076
Phone: (512)471-5602
Fax: (512)471-0592

Thomas T. Stocksdale
S.C. Johnson & Son, Inc.
1525 Howe Street
Racine, WI 53403
Phone: (414)631-2000
Fax: (414)631-2167

6.4. CONTAMINATION SUBJECT TO TREATMENT

Table 6.1 lists the nine most important chlorinated alkanes in ground water in a survey of 358 hazardous wastes sites (Plumb and Pitchford, 1985). Vinyl chloride in ground water is almost invariably produced from the reductive dechlorination of tetrachloroethylene or trichloroethylene. The other compounds are used as solvents and find their way into ground water through improper disposal.

TABLE 6.1.   THE COMMON CHLORINATED ORGANIC COMPOUNDS OCCURRING AS
CONTAMINANTS OF GROUND WATER

Compound	Detection frequency (% of 358 sites)	Average concentration in all samples (mg/l)	Average concentration in samples where detected (mg/l)
Trichloroethylene	51	2.0	3.8
Tetrachloroethylene	36	3.5	9.7
Chloroform	28	0.41	not ranked
Methylene Chloride	19	2.2	11.2
1,1,1-Trichloroethane	19	0.24	not ranked
1,2-Dichloroethane	14	0.90	6.3
Vinyl Chloride	8.7	not ranked	not ranked
Chlorobenzene	not ranked[a]	not ranked	not ranked
Carbon Tetrachloride	not ranked	not ranked	not ranked

[a] "Not ranked" means not ranked in the top twenty organic contaminants at the 358 sites surveyed.

Trichloroethylene, chloroform, 1,1,1,-trichloroethane, 1,2-dichloroethylene, and dichloromethane (methylene chloride) can be biologically cooxidized during growth on a variety of substrates (methane, propane, toluene) that exist as vapors and can be delivered to the subsurface environment through the flow of air (Wilson and White, 1986; Henson et al., 1988, Wackett and Gibson, 1988). Chlorobenzene, dichloromethane, 1,2-dichloroethane, and vinyl chloride can also serve as primary substrates for microbial growth (Janssen et al., 1985; Davis and Carpenter, 1990; Rittmann and McCarty, 1980); they do not necessarily require a cosubstrate for biodegradation, although removal in the presence of a cosubstrate may be more rapid. There is no known pathway for aerobic biodegradation of tetrachloroethylene or carbon tetrachloride.

With the exception of trichloroethylene and vinyl chloride, the average concentrations listed in Table 6.1 can be easily tolerated by heterotrophic bacteria. Cooxidation of trichloroethylene and vinyl chloride can occur through an epoxide that is chemically reactive, and is considerably more toxic than the parent compound.

Trichloroethylene concentrations below 3 mg/l air do not inhibit the rate of oxidation of the primary substrate (Table 6.2). Less is known about toxic effects of vinyl chloride; concentrations of 1.0 mg/l air are generally tolerated in laboratory microcosms.

TABLE 6.2.   EFFECT OF THE CONCENTRATION OF TRICHLOROETHYLENE ON THE RATE OF BIODEGRADATION OF AVIATION GASOLINE[a] VAPORS IN SOIL MICROCOSMS

Concentration of Trichloroethylene		Degradation of Gasoline Vapors
(mg/kg soil)	*(mg/l air, if all trichloroethylene volatilized)*	*(mg/kg day)*
45,600		0.06
4,500	1,000	0.11
13	3	24
4.1	0.9	26
1.2	0.3	31
0.0	0.0	33

a The hydrocarbons in gasoline support the cooxidation of trichloroethylene.

6.5.   SPECIAL REQUIREMENTS FOR SITE CHARACTERIZATION

Sites should be carefully mapped to separate areas with oily phase liquids from areas that only have contamination in air or water. Areas containing oily phase liquids have a great capacity to contaminate soil air or ground water. In regions containing oily phase liquids, the mass of contaminant removed through biological treatment is trivial compared to the mass of contaminant that will partition into air or water and be carried away. Air or water should not be injected into geological material containing oils unless there is an intent to recover the effluent through pump and treat or soil vacuum extraction.

Many contaminated subsurface materials contain chlorinated solvents sorbed to organic materials. These materials can act as source areas for contamination of ground water or soil air and are good candidates for in-situ bioventing. Core material should be extracted to determine the total mass of sorbed chlorinated solvents that are subject to remediation. Desorption isotherms can be useful to determine the extent of remediation required to meet cleanup goals for ground water or soil air.

6.6.　SITE CHARACTERISTICS THAT ARE PARTICULARLY FAVORABLE

Transmissive material is better than less transmissive material, because injection and extraction wells can be spaced on wider intervals, and the power requirement for blowers is less. A deep water table provides more volume of unsaturated material, and thus more residence time on a surface area basis, for treatment of vapors.

Direct metabolism or cooxidation of chlorinated solvents ultimately produces hydrochloric acid. Unless the geological matrix contains carbonates or other natural buffers, the pH will drop to levels that inhibit oxidation of the primary substrate. If ground-water pH is controlled by a carbonate/bicarbonate buffering system sustained by a source of carbonate in the aquifer matrix, biological activity can proceed for much longer periods of time.

In-situ treatment need not impede the beneficial use of infrastructure at a site. Sites with infrastructure of great economic value (such as refineries or factories) or sites of great importance to health and safety (such as highways, hospitals, certain military installations) are good candidates for in-situ remediation. In-situ remediation is also particularly appropriate if contamination has migrated to property owned by a second party.

6.7.　SITE CHARACTERISTICS THAT ARE PARTICULARLY UNFAVORABLE

Material that contains secondary porosity (such as cracks, channels or fissures) allows short circuiting of gases. These secondary passages will form if the material has enough clay or organic matter to form water-stable aggregates.

The primary substrates are hydrocarbons supplied at concentrations in air that are near or within the explosive range. The potential for migration of explosive vapors in basements, sewers, utility conduits, and other underground excavations should be assessed.

Material that has wide variations in texture is difficult to treat with any technology that circulates fluids. Above the water table, fine textured materials tend to retain organic liquids by capillary attraction. Because fine textured materials have more clay and organic matter, they also have a greater sorptive capacity. Remedial fluids tend to flow around rather than through the most contaminated material.

Sites with unfavorable characteristics that are being remediated prior to sale or transfer to a second party may not be remediated in an acceptable time frame.

6.8.   PERFORMANCE UNDER OPTIMAL CONDITIONS

6.8.1. Importance of the Rate Law

The active microbial populations in unsaturated soil behave as if they are contained in a thin film of water in chemical equilibrium with soil air and ground water. At high concentrations, the enzymatic machinery of the organisms degrading the compound are saturated. The rate of degradation is proportional to the amount of active biomass, but is independent of the concentration of organic compound. Disappearance of the organic contaminant is described with zero-order kinetics, where the decrease in contaminant mass is linear with time. The rate of biodegradation is most conveniently normalized to the dry weight of soil in contact with contaminated soil air or ground water.

At low concentrations, the supply of organic compound is limiting, and the rate of degradation is proportional to the concentration of the organic compound in contact with the active microorganisms as well as the amount of active biomass. Disappearance of the organic compound is described by first-order kinetics expressed as a half-life, or by a first-order rate constant.

6.8.2. Importance of Partitioning

Interpretation of first-order rate laws, and particularly extrapolation of experimental data to other systems, is complicated by partitioning of the organic compound between soil air, water and solids. Laboratory experiments done with columns usually have a natural ratio of air, water, and solids; but batch laboratory experiments usually do not. If the batch laboratory systems have a greater proportion of air, the kinetics of depletion in the laboratory system will be slowed with respect to the depletion that would be seen at field scale. The factor by which kinetics are slowed can be estimated by dividing the ratio of air to solids in the experimental system by the ratio of air to solids in the natural system. In Table 6.3, simple chemodynamic theory was used to predict the partitioning of the chlorinated solvents between air, water, and solids in a representative subsurface material. For most compounds, the majority of contaminant mass is in the soil water and therefore available to microorganisms. However, biodegradation of vinyl chloride will be strongly influenced by partitioning, particularly in cold subsurface material.

After acclimation, the biological removal of vapors of natural hydrocarbons is very rapid. Typical vadose zone subsurface material, as depicted in Table 6.3, has at most 100 ml of air per kg of soil. The air could contain at most 25 mg of oxygen gas, which would support metabolism of approximately 7 or 8 mg/kg of hydrocarbon. Table 6.4 presents the rate of hydrocarbon oxidation in fertile, well acclimated subsurface material. Oxygen would be exhausted in a few hours in these soils at typical air content.

Table 6.5 presents laboratory data on the kinetics of mineralization of trichloroethylene, chloroform, and 1,2 dichloroethane vapors. To provide a common basis for comparison, all rates were calculated as zero-order rates normalized to the mass of soil. In general, the rate of removal of chlorinated solvents was one to ten percent of the removal of natural hydrocarbons.

TABLE 6.3.   PARTITIONING OF CHLORINATED ORGANIC COMPOUNDS BETWEEN AIR, WATER, AND SOLIDS IN A HYPOTHETICAL SUBSURFACE MATERIAL WITH AN AIR-FILLED POROSITY OF 0.2, A WATER-FILLED POROSITY OF 0.2, AND AN ORGANIC CARBON CONTENT OF 100 MG/KG

Compound	Air Percent of total	Water Percent of total	Solids Percent of total
Trichloroethylene	20	72	7.3
Tetrachloroethylene	20	62	18
Chloroform	7.0	90	3.1
Methylene Chloride	6.2	93	0.7
1,1,1-Trichloroethane	24	60	15
1,2-Dichloroethane	2.3	97	1.1
Vinyl Chloride 25°C	58	42	0.08
Vinyl Chloride 10°C	96	4.0	trivial
Chlorobenzene	7.2	66	27
Carbon Tetrachloride	17	29	53
Cosubstrates			
Toluene	12	78	10
Methane	95	5	trivial

TABLE 6.4.   KINETICS OF DEPLETION OF NATURAL HYDROCARBONS IN UNSATURATED SOIL AND SUBSURFACE MATERIAL

Soil type (Ref)	Methane	Propane and Butanes	Benzene	Toluene	Ethyl-benzene	o-Xylene
		---------------------------------- (mg/kg soil/day) ----------------------------------				
Sand (a)			43.6	10.1	3.6	2.5
Sand (a)			19.6	4.0	2.0	0.8
Loam (a)			19.3	4.4	1.8	1.1
Sandy Loam (b)			319			146
Loamy Clay (c)	104					
Sandy Silty Loam (c)	157					
Sandy Clay Loam (c)	173					
Sand (d)	1,400					
No Data (e)	150			288		
Muck (f)		163				

a. Miller and Canter, 1991; b. English, 1991; c. Bender and Conrad, 1992; d. Hoeks, 1972; e. Anonymous; f. Kampbell et al., 1987

The hypothetical representative subsurface material described in Table 6.3 has 0.1 l of soil water per kg. If the data in Table 6.1 on typical concentrations of chlorinated solvents in ground water are divided through by ten, they can be used to estimate the mass of contaminant in mg per kg of typical aquifer material. Typical concentrations of chlorinated solvents are near 1 mg/kg. The rates of removal of chlorinated hydrocarbon presented in Tables 6.5 and 6.6 indicate that the major fraction of chlorinated solvent contamination could be removed in a few hours to a few days.

TABLE 6.5.   MINERALIZATION OF VAPORS OF CHLORINATED SOLVENTS IN SOIL ACCLIMATED TO DEGRADE VAPORS OF NATURAL HYDROCARBONS

Refer-ence	Soil Type	Substrate and Nutrients	Initial Solvent Conc.		Mineralization Rate (mg/kg soil/day)	
			mg/kg soil	μg/l air	With Substrate	Without Substrate
Trichlororethylene						
(a)	Sandy clay	2.5% methane N,P,K,S	50	20,000	0.39	0.35
(b)	Organic rich muck	0.2% propane butanes	3.2	1,100	27	
(c)	no data	N,P,S		1,030		<10
(c)	no data	0.1% toluene N,P,S		1,030	108	
(d)	Coarse sand	2.0% methane applied 4 times	1.5	260	0.06	<0.006
Chloroform						
(a)	Sandy clay	2.3% methane N,P,K,S	100	40,000	0.19	0.08
1,2-Dichloromethane						
(a)	Sand	1.3% methane N,P,S,K	100	40,000	0.71	0.83
(a)	Silty loam	2.2% methane N,P,S,K	100	40,000	0.12	0.22
(a)	Sandy clay	1.1% methane N,P,S,K	100	40,000	1.39	1.31
(a)	Sandy clay	4.9% methane	100	40,000	0.06	0.04

a. Speitel and Closmann, 1991; b. Kampbell et al., 1987; c. Anonymous; d. Broholm et al, 1991.

TABLE 6.6. **REMOVAL OF VAPORS OF TRICHLOROETHYLENE AND VINYL CHLORIDE IN SUBSURFACE MATERIAL UNDER OPTIMAL CONDITIONS**[a]

Source	Soil Type	Substrate and Nutrients	Initial Solvent Conc.		Mineralization Rate (mg/kg soil/day)
			mg/kg soil	μg/l air	
Trichloroethylene					
Tucson, Arizona	Sandy	0.6% gasoline vapors N,P,K	17	4,200	1.47
St. Joseph, Michigan	Sand	0.6% gasoline vapors N,P,K	17	4,200	2.00
Racine, Wisconsin	Organic Rich Loam	0.6% gasoline vapors N,P,K	17	4,200	0.53
Vinyl Chloride Racine, Wisconsin	Loam	0.6% gasoline vapors	4.0	1,000	1.75
Tucson, Arizona	Sand	0.6% gasoline vapors N,P,K	4.0	1,000	1.35

[a] Kampbell and Wilson, 1993.

6.9. PROBLEMS ENCOUNTERED WITH THE TECHNOLOGY

If concentrations are high, in the milligrams per liter range, bioventing can economically remove a major fraction of contaminant mass prior to polishing with activated carbon to meet regulatory endpoints. If concentrations are low, within one or perhaps two orders of magnitude of the goal, bioventing can meet acceptable concentrations with a relatively few applications of substrate or oxygen. However, bioventing to degrade chlorinated organics through cooxidation should not be expected to reduce the average concentrations presented in Table 6.1 down to drinking water MCLs (Maximum Contaminant Levels).

6.10. RELEVANT EXPERIENCE WITH SYSTEM DESIGN

S.C. Johnson & Son, Inc., operates a facility in Racine, Wisconsin, that packages wax products in aerosol cans. A mixture of propane and butanes is used as the propellant gas. To treat propellant gas that escapes during the filling process, they constructed a soil bed bioreactor that is 184 feet long, 159 feet wide and 4 feet deep. Air

containing the propane and butanes is injected at the bottom of the bed, and works its way
to the surface (Figure 6.2). The waste stream at their facility is very similar to air that
would be used for the intentional cooxidation of chlorinated solvents. Soil acclimated to
degrade propellant gas also degrades trichloroethylene. The performance of their soil
bed reactor is the best model available for bioventing of chlorinated solvents at field scale.

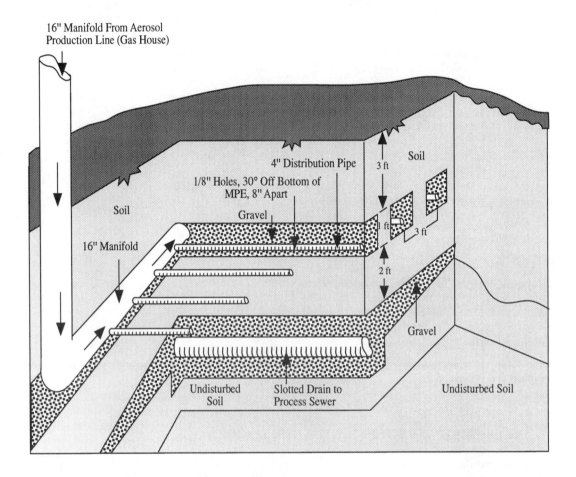

Figure 6.2. Soil bed constructed by S.C. Johnson & Son, Inc. in Racine, Wisconsin, to treat an
airstream containing 2,000 to 3,500 ppm of propellant gas (a mixture of propane and
butanes).

The performance of the soil bed reactor was related to the air loading rate. The
removal efficiency for propellant gas is greater than 90% when the flow of air is less than
3 cubic feet per minute per 100 square feet (cfm/100 sq. feet) of bed surface area. Removal
drops to less than 50% at flows greater than 6 cfm/100 sq. feet (personal communication,
Thomas T. Stocksdale, Safety and Environmental Affairs Manager, S.C. Johnson & Son,
Inc., Racine, Wisconsin, see Figure 6.3 for data). At a bed depth of four feet, and an
estimated air-filled porosity of 0.2, 3 scf/100 sq. feet corresponds to an average residence
time for air of 30 minutes.

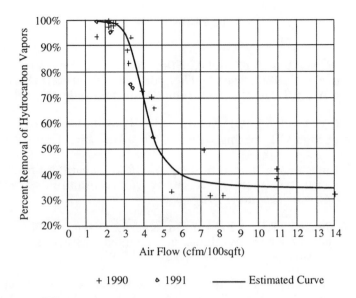

Figure 6.3.   **Effect of flow rate on removal efficiency of propellant gas hydrocarbons in a soil bed bioreactor.**

If this soil bioreactor were used to treat air containing 1,100 μg/l of trichloroethylene, 0.22 mg of trichlorethylene vapor would be exposed to each kg of soil for 30 minutes. If it performed according to the laboratory study of Kampbell et al. (1987, see Table 6.5), this reactor could remove 0.56 mg/kg of trichloroethylene in 30 minutes under zero-order kinetics. If this behavior is general, subsurface material optimized to remove the primary substrate will remove vapors of chlorinated solvents to low concentrations where first-order kinetics apply.

There was extensive lateral migration of air injected into their bed. To prevent migration of injected air under their facility, they installed sheet piling along three sides of the soil bed. "Hot spots" developed in their bed that had high flow rates of air and poor removal of propellant gas hydrocarbons. To correct "hot spots" they mechanically deep till and regrade the soil.

REFERENCES

Anonymous.   Roy F. Weston, Inc.   1990. *Final Report Task Order 8: Biotreatment of Gaseous-Phase Volatile Organic Compounds.* Contract DAAA 15-88-D-0010. United States Army Toxic and Hazardous Materials Agency.

Bender, M., and R. Conrad. 1992. Kinetics of CH_4 oxidation in oxic soils exposed to ambient air or high CH_4 mixing ratios. FEMS Microbiol. Ecol. 101(1992):261-270.

Broholm, K., T.H. Christensen, and B.K. Jensen. 1991. Laboratory feasibility studies on biological in-situ treatment of a sandy soil contaminated with chlorinated aliphatics. *Environ. Technol.* 12:279-289.

Broholm, K., B.K. Jensen, T.H. Christensen. and L. Olsen. 1990. Toxicity of 1,1,1-trichloroethane and trichloroethene on a mixed culture of methane-oxidizing bacteria. *Appl. Environ. Microbiol.* 56(8):2488-2493.

Davis, J.W., and C.L. Carpenter. 1990. Aerobic biodegradation of vinyl chloride in groundwater samples. *Appl. Environ. Microbiol.* 56(12):3868-3880.

English, C.W. 1991. *Removal of Organic Vapors in Unsaturated Soil.* Dissertation presented to the Faculty of the Graduate School of the University of Texas at Austin. May 1991.

Henson, J.M, M.V. Yates, J.W. Cochran, and D.L. Shackleford. 1988. Microbial removal of halogenated methanes, ethanes, and ethylenes in an aerobic soil exposed to methane. *FEMS Microbiol. Ecol.* 52(3-4):193-201.

Hoeks, J. 1972. Changes in composition of soil air near leaks in natural gas mains. *Soil Science.* 113:46.

Janssen, D.B., A. Scheper, L.Dijkhuizen, and B. Witholt. 1985. Degradation of halogenated aliphatic compounds by *Xanthobacter autotrophicus* GJ10. *Appl. Environ. Microbiol.* 49(2):673-677.

Kampbell, D.H. and B.H. Wilson. 1993. Bioremediation of chlorinated solvents in the vadose zone. In: *In Situ and On-Site Bioreclamation. The Second International Symposium.* April 5-8, 1993. San Diego, California. In Press.

Kampbell, D.H., J.T. Wilson, H.W. Read, and T.T. Stocksdale. 1987. Removal of volatile aliphatic hydrocarbons in a soil bioreactor. *J. Air Pollut. Cont. Assoc.* 37(10):1236-1240.

Miller, D.E., and L.W. Canter. 1991. Control of aromatic waste air streams by soil bioreactors. *Environ. Progress.* 10(4):300-306.

Plumb, R.H., and A.M. Pitchford. 1985. Volatile organic scans: Implications for ground water monitoring. In: *Proceedings of The Petroleum Hydrocarbons and Organic Chemicals in Ground Water - Prevention, Detection, and Restoration.* National Water Well Association. Dublin, OH. pp. 207-221.

Rittmann, B.E., and P.L. McCarty. 1980. Utilization of dichloromethane by suspended and fixed-film bacteria. *Appl. Environ. Microbiol*. 39(6):1225-1226.

Speitel, G.E., and F.B. Closmann. 1991. Chlorinated solvent biodegradation by methanotrophs in unsaturated soils. *J. Environ. Eng*. 117(5):541-558.

Wackett, L.P., and D.T. Gibson. 1988. Degradation of trichloroethylene by toluene dioxygenase in whole-cell studies with *Pseudomonas putida* F1. *Appl. Env. Microbiol*. 54(7):1703-1708.

Wilson, B.H., and M.V. White. 1986. A fixed-film bioreactor to treat trichloroethylene-laden waters from interdiction wells. In: *Proceedings of The Sixth National Symposium and Exposition on Aquifer Restoration and Groundwater Monitoring*. National Water Well Association. Dublin, OH. pp. 425-435.

SECTION 7

IN-SITU BIOREMEDIATION TECHNOLOGIES FOR PETROLEUM-DERIVED HYDROCARBONS BASED ON ALTERNATE ELECTRON ACCEPTORS (OTHER THAN MOLECULAR OXYGEN)

Martin Reinhard
Stanford University
Department of Civil Engineering
Stanford, California 94305-4020
Telephone: (415)723-0308
Fax: (415)725-8662

7.1. FUNDAMENTAL PRINCIPLES

As currently practiced, conventional in-situ biorestoration of petroleum-contaminated soils, aquifer solids, and ground water relies on the supply of oxygen to the subsurface to enhance natural aerobic processes to remediate the contaminants. However, anaerobic microbial processes can be significant in oxygen-depleted subsurface environments that are contaminated with petroleum-based compounds and/or chlorinated solvents. The purpose of this chapter is to discuss anaerobic biotransformation of petroleum-derived ground-water contaminants, to discuss both laboratory and field evaluation of the process, and to discuss important site conditions that would influence a successful bioremediation.

7.1.1. Comparison of Oxygen and Alternate Electron Acceptor Based In-Situ Bioremediation Technologies

In-situ bioremediation technology for the decontamination of soil and ground water contaminated with petroleum-derived hydrocarbons involves the stimulation of naturally occurring microorganisms that are capable of degrading the organic contaminants (Atlas, 1981; Lee et al., 1988). Biostimulation consists of adding those nutrients and/or electron acceptors that limit bacterial growth to the contaminated zone. Bioaugmentation, on the other hand, involves introduction of adapted or genetically engineered microorganisms into the contaminated aquifer. Using contaminants as substrates for energy and growth, microorganisms convert the contaminants into harmless products, principally CO_2, cell mass, inorganic salts, and water. When oxygen is consumed, anaerobic microorganisms may grow using alternate electron acceptors.

Anaerobic degradation of aromatic hydrocarbons has initially been identified at field sites (Reinhard et al. 1984) and in microcosm studies (Wilson et al., 1987) and has now been demonstrated in the laboratory under a number of redox conditions, including reduction of nitrate, iron(III) and manganese(IV) oxides, sulfate, and carbon dioxide. In contrast to aromatic hydrocarbons, aliphatic hydrocarbon degradation without oxygen has not been reported. In aquifers contaminated with biodegradable organic compounds, electron acceptors tend to be used successively in order of decreasing free energy yield. Oxygen is the preferred electron acceptor, followed by nitrate, manganese(IV) and iron(III) oxides (MnO_2 and $FeOOH$, respectively), sulfate, and carbon dioxide. This sequence applies to pH 7 and should be valid for most field conditions where the appropriate microorganisms occur.

The conventional approach to hydrocarbon bioremediation is based on aerobic processes. Anaerobic bioremediation has been tested only in a very few cases and is still considered experimental. For instance, in a review of 17 sites contaminated with hydrocarbon fuels and oils (Staps, 1990), hydrogen peroxide was used as the electron acceptor at seven sites, air at five, combinations of nitrate-ozone and nitrate-air at one site each, and nitrate alone was used only at three sites. Much available information has been developed in laboratory studies; however, the applicability of these results to field conditions remains to be studied. Anaerobic transformation rates can be slow and lag times long and unpredictable, except for transformation in denitrifying systems which can be fast. In spite of slow rates, anaerobic bioremediation could play a significant role in the future mainly because the principal factor limiting aerobic bioremediation, the difficulty of supplying oxygen to the subsurface, is circumvented.

Although aerobic biodegradation of refined petroleum products is relatively rapid and complete under ideal growth conditions, application of anaerobic processes may be preferable in ground waters because ideal aerobic growth conditions are difficult to maintain in an aquifer. Rapid aerobic degradation requires ample supply of nutrients and oxygen, good mixing, and a high microbial mass, conditions that are difficult to maintain in aquifers (Wilson, B. et al., 1986; Lee et al., 1988). Furthermore, at many sites there may be a very high abiotic oxygen demand due to hydrogen sulfide, Fe^{2+} or other readily oxidizable compounds, making it difficult to increase the reduction potential into the aerobic range (> 0.82 V, see Table 8.5, Section 8). The advantages of aerobic bioremediation may become inconsequential if the overall degradation rate is controlled by slow dissolution, dispersion and/or desorption. Mass transfer limitations are especially severe at sites where petroleum-derived hydrocarbons are present as nonaqueous phase liquids (NAPLs). Under such conditions, natural or passive biological degradation (aerobic or anaerobic) may be sufficiently fast for removing hydrocarbons that are slowly released into the ground water.

At contaminated sites, a range of other, site-specific factors can limit biotransformation, such as the occurrence of metals or other toxics, accumulation of toxic intermediates and suboptimal temperatures. Nearby NAPL concentrations may reach toxic levels, thereby limiting biological activity. It is unknown whether aerobic or anaerobic processes are more readily inhibited by such factors. Thus, the decision to use either aerobic or anaerobic processes may depend on site-specific conditions.

Since the consumption of O_2 is relatively fast and the rate of O_2 supply is slow due to low O_2 solubility in water, expansion of the aerobic zone is limited by the rate of O_2 supply to the aquifer. Anaerobic conditions are expected to persist within aerobically treated aquifers, especially in relatively impermeable zones and zones further away from the injection wells. Water is an inefficient mass transfer medium for O_2 due to the low water solubility of O_2. For the degradation of relatively small amounts of hydrocarbons, large amounts of water need to come in contact with the aquifer solids. The complete oxidation of 1 mg hydrocarbon compounds requires 3.1 mg O_2 (Hutchins and Wilson, 1991). Thus, for the bioremediation of 1 kg of aquifer material containing 10 g/kg hydrocarbon compounds, a minimum of 3.1 m^3 of oxygenated water containing 10 mg/l O_2 must be supplied. Potentially, the overall degradation efficiency can be increased by using alternate electron acceptors that are more water soluble. The ratio of feed water to contaminant mass degraded is higher if the electron acceptor concentration in the feed is increased. Nitrate salts are much more water soluble (92 g/l or 1.33 M as sodium nitrate) than O_2 (10 mg/l or 0.31 mM). Comparing the water solubilities and the half-reactions for O_2 and nitrate reduction (Table 8.5, Section 8),

$$2NO_3^- + 12H^+ + 10\,e^- \longrightarrow N_2 + 6H_2O \tag{1}$$

$$O_2 + 4H^+ + 4e^- \longrightarrow 2H_2O \tag{2}$$

it is evident that the reducing equivalents that can be introduced into an aquifer using saturated sodium nitrate solution is approximately 50 times higher than with a saturated oxygen solution.

The use and potential benefits of anaerobic in-situ remediation technology are the subject of this review. Nitrate, sulfate, iron(III) oxide and carbon dioxide and, to a very limited extent, mixed electron acceptor systems are considered. However, only nitrate and nitrate in combination with oxygen sources have been tested in field applications.

The general approach to in-situ bioremediation for ground-water cleanup has been summarized by Sims et al. (1992). It is applicable to both aerobic and anaerobic processes and consists of the following basic tasks:

(1)   **Site investigation** to determine the distribution, mobility and fate of the contaminants under site specific conditions.

(2)   Performance of **treatability studies to determine the potential for bioremediation** and to define the required operating and management practices.

(3)   Development of a **bioremediation plan** based on fundamental engineering principles.

(4)   Establishment of a **monitoring program** to evaluate performance of the remediation effort.

The principal design considerations of the aerobic and anaerobic bioremediation processes are the same because water serves as the nutrient feed solution in both cases. The principal factors that must be considered include all aspects affecting mass transfer, substrate retardation, and bioavailability. For anaerobic bioremediation, special attention must be given to the adaptation status and growth condition of the indigenous anaerobic bacteria, and the initial redox status of the aquifer. Potential advantages of anaerobic bioremediation include:

(1)   Alternate electron acceptors (except ferric iron) are more water soluble and, consequently, require lower volumes of nutrient solution to be supplied to the contaminated zone.

(2)   Reduced plugging problems because of lower biomass yields of anaerobic bacteria and a lesser tendency for iron precipitation.

7.1.2.   Hydrocarbon Transformation Based on Alternate Electron Acceptors

7.1.2.1.   Laboratory Studies

Grbic-Galic (1989, 1990) summarized the laboratory research on anaerobic hydrocarbon transformation involving microcosms and enrichment cultures. Most laboratory studies have attempted to (1) demonstrate biotransformation and mineralization of the substrate by obtaining carbon mass balances, (2) identify the electron acceptor by determination of the reaction stoichiometry, and (3) determine optimal growth conditions. Most studies have been conducted with microcosms or enrichment cultures; although, in a few cases, isolation of pure strains has been reported. For example, Dolfing et al. (1990) and Lovley and Lonergan (1990), respectively, reported isolation of pure nitrate- and iron(III)-reducing strains capable of using aromatic hydrocarbons as a sole carbon source. However, growth parameters and the effect of environmental conditions on aromatic transformation have been investigated only in a very few studies.

Denitrifying Systems

Under denitrifying conditions, oxidation of monoaromatic compounds has been demonstrated in a number of systems (e.g., Kuhn et al., 1988; Mihelcic and Luthy, 1988; Altenschmidt and Fuchs, 1991; Ball et al., 1991; Evans et al., 1991a; Evans et al., 1991b; Flyvbjerg et al., 1991; Hutchins et al., 1991a; Evans et al., 1992). The stoichiometry of the denitrification reaction of toluene assuming no cell growth with NO_3^- reduced completely to N_2 and toluene completely oxidized to CO_2 is

$$C_7H_8 + 7.2 H^+ + 7.2 NO_3^- \longrightarrow 3.6 N_2 + 7.6 H_2O + 7 CO_2 \qquad (3)$$

When biodegradation of benzene, toluene, ethylbenzene and xylenes (BTEX) mixtures was tested under denitrifying conditions, degradation tended to be sequential, with toluene being the first substrate to be degraded, followed by the degradation of p- and m-xylene, ethylbenzene and o-xylene. Mihelcic and Luthy (1988) reported the degradation of naphthalene. Benzene does not seem to be degraded (Kuhn et al., 1988; Evans et al., 1991a; Evans et al., 1991b; Hutchins et al., 1991a) although in one study Major et al. (1988) reported removal under conditions thought to be denitrifying. Hutchins et al. (1991a) reported longer lag times and slower degradation rates in core material contaminated with JP-4 aviation fuel than in uncontaminated core material.

Using an enrichment culture and ethylbenzene as the substrate, Ball et al. (1991) have shown that single aromatic substrates can be degraded rapidly (within hours) and that nitrate reduction to nitrogen gas proceeds through nitrite. Similar findings were reported by Evans et al. (1991a, 1991b) for toluene. Ball et al. (1991) also demonstrated that composition and preparation of the growth medium can affect the observed transformation rates.

Ball et al. (1991) tested inocula from different sources for the potential to degrade BTEX compounds. They found that microorganisms with the ability to degrade aromatic hydrocarbons are not ubiquitous. Sewage seed that contains a diverse population of microorganisms, for instance, did not adapt to the aromatic compounds tested. Much work remains to be done before such experimental data can be interpreted with confidence and the biodegradation potential under field conditions can be predicted.

Sulfate-Reducing Systems

Bioremediation using sulfate as the electron acceptor involves oxidation of aromatic hydrocarbons by sulfidogenic organisms coupled with reduction of sulfate to hydrogen sulfide (Edwards et al., 1991; Haag et al., 1991; Beller et al., 1992; Edwards et al., 1992). For toluene, the stoichiometry of this reaction, assuming no cell growth, may be written as (Beller et al., 1992):

$$C_7H_8 + 4.5 SO_4^{2-} + 3 H_2O \longrightarrow$$
$$2.25 HS^- + 2.25 H_2S + 7 HCO_3^- + 0.25 H^+ \qquad (4)$$

As in some denitrifying systems, degradation under sulfate-reducing conditions is also sequential with toluene being the preferred substrate, followed by p-xylene and with o-xylene degraded last (Edwards et al., 1991, 1992). Ethylbenzene and benzene were not degraded under the conditions of the experiment. In a follow-up study, Edwards and Grbic-Galic (1992) observed benzene degradation in the absence of all other aromatic substrates. After a lag time of 30 days under strictly anaerobic conditions, these authors observed mineralization of benzene and suspected sulfate to be the electron acceptor. Accumulation of HS⁻ may inhibit the process, however, and is a problem that remains to be resolved.

Iron(III)-Reducing Systems

Lovley and Lonergan (1990) have isolated an iron-reducing bacterium capable of degrading toluene, p-cresol and phenol. For toluene, the stoichiometry of the process is

$$C_7H_8 + 36\ Fe^{3+} + 21\ H_2O \longrightarrow 36\ Fe^{2+} + 7\ HCO_3^- + 43\ H^+ \tag{5}$$

whereby 36 moles of Fe(III) are required to oxidize one mole of toluene. Relative to other anaerobic processes, Fe(III) reduction has a very unfavorable substrate to electron acceptor ratio. Lovley et al. (1989) found that toluene was transformed into CO_2 and Fe^{2+} at a ratio which agreed with the above stoichiometry.

Transport of the dissolved Fe(II) from the aquifer could cause secondary problems such as clogging and fouling of the aquifer. Furthermore, the supply of large amounts of colloidal iron(III) oxide or soluble Fe(III) citrate (Lovley et al., 1989) to an aquifer has not been tested. To develop bioremediation strategies based on iron reduction, a better understanding of occurrence, nutritional requirements, growth conditions and metabolism of iron-reducing bacteria must be developed.

Fermentative/Carbon Dioxide-Reducing Systems

Under methanogenic/fermentative conditions, several aromatic hydrocarbon compounds, including benzene and toluene, have been shown to transform into CO_2 and methane (Grbic-Galic and Vogel, 1987). The culture originated from sewage seed and was enriched under methanogenic conditions using ferulic acid as the only carbon source. Assuming no cell growth, the stoichiometry for the transformation reaction is

$$C_7H_8 + 5\ H_2O \longrightarrow 2.5\ CO_2 + 4.5\ CH_4 \tag{6}$$

Biotransformation was studied with toluene or benzene as the only carbon source. Biotransformation began after a three month lag time and was complete after 60 days of incubation. Since this ground-breaking study, several other aromatic substrates have been shown to be degraded under methanogenic conditions, including styrene, naphthalene and acenaphthalene (Grbic-Galic, 1990), as well as benzothiophene, a sulfur-containing heterocyclic compound (Godsy and Grbic-Galic, 1989).

Fermentation/methanogenic degradation could be used as a passive bioremediation technology (Section 9) and is likely to be an ongoing process at many sites where the geochemical conditions have evolved naturally, without human intervention. Reliable assessment of the process is difficult under field conditions since mass balances are difficult to establish. An indication that the process is occurring is the detection of methane in combination with characteristic intermediates such as aromatic acids (Reinhard et al., 1984; Wilson et al., 1987; Baedecker and Cozzarelli, 1991).

Mixed Electron Acceptor Systems

In aquifer segments augmented by electron acceptors, different electron acceptors are likely to co-occur either within the same aquifer compartment, or spatially separated into adjacent compartments. Few laboratory studies have examined mixed electron acceptor systems, although they are likely to be common at field sites. For instance, at the sites where denitrifying conditions were investigated, O_2 was frequently present in the nitrate feed water. Both electron acceptors were consumed, but the effect of the oxygen on the overall process was not determined.

Different electron acceptors and products of aromatic degradation processes can react with each other in a number of biological and chemical reactions. Beller et al. (1992) have studied the link between dissimilatory sulfate reduction to sulfide and iron(III) reduction to iron(II) by a sulfate-reducing enrichment culture. Ferric iron appeared to reoxidize hydrogen sulfide in an abiotic process and/or lower the inhibitory effect of hydrogen sulfide. Toluene was the sole carbon and energy source, but other substrates were not tested.

7.1.2.2. Large Scale Bioremediation Studies Using Nitrate

Nitrate is the only alternate electron acceptor with demonstrated potential for use in large scale in-situ bioremediation applications involving petroleum-derived hydrocarbons (Table 7.1). The other possible alternate electron acceptors (iron(III), sulfate, and CO_2) have been found in systems that may be classified as passive bioremediation, such as in landfill leachate plumes (Reinhard et al., 1984) and at spill sites (Lovley et al., 1989).

Table 7.1 summarizes results of selected field studies, where denitrification was tested as a means to remove aromatic hydrocarbon contamination. The Traverse City and the Rhine River Valley studies involved actual contamination sites. The Borden experiment involved injection into the aquifer of a mixture containing benzene, toluene, and xylene isomers (BTX) in one experiment and gasoline in another. In general, BTX compounds were found to disappear within the nitrate amended zone. Interpretation of these data, however, was complicated by a number of factors, especially the co-occurrence of nitrate and O_2, and the lack of complete characterization of the organic substrates. Nitrate removal exceeded the expected amount based on the substrates analyzed, and this was attributed to the dissolved organic carbon in ground water (Berry-Spark et al., 1988). Werner (1985) proposed that if O_2 and nitrate are present simultaneously, O_2 is used for the first oxidation step to produce partially oxygenated products and nitrate is then used for mineralization of the oxidation products.

7.1.3. Maturity of the Technology

Anaerobic transformation of aromatic hydrocarbon compounds is a very recent discovery, and it is too early to predict an impact on bioremediation technology. The technology to use nitrate has been tested in a limited number of cases and is still highly immature, although available data are very promising. Research on aromatic transformation under all other reducing conditions is still at very early stages. Currently, it is not possible to predict the site conditions under which biostimulation using nitrate or sulfate will be successful and define optimal operating conditions for a given site.

TABLE 7.1.   FIELD STUDIES WHERE DENITRIFICATION HAS BEEN EVALUATED

STUDY SITE AND AUTHORS	CONTAMINATION AND CONDITIONS	MAJOR IMPLICATION FOR IN-SITU BIOREMEDIATION
Traverse City, MI, Hutchins and Wilson, 1991	JP-4 fuel; $NaNO_3$: 62 mg/l; O_2: 0.5 to 1 mg/l	(1) removal of benzene, toluene, *m,p*-xylene; recalcitrance of *o*-xylene. (2) nitrate removed exceeded stoichiometric amount of BTEX removal. (3) partitioning of compounds into the water phase appears to be a major factor determining compound removal.
Borden, Ontario, Berry-Spark et al., 1988	Gasoline and BTX; Oxygen and nitrate	(1) BTX transform more slowly when gasoline is present than in systems where BTX are the only substrate. (2) In systems containing both O_2 and nitrate, aerobic and (facultative) denitrifying organisms appear to cooperate.
Seal Beach, Reinhard et al., 1991	Gasoline contaminated ground water feed; NO_3^- (6 mg/l)	90% nitrate removal in mixed nitrate/sulfate system, aromatics removal toluene>*p*-xylene>*o*-xylene>benzene
Rhine Valley, FRG, Werner, 1985	Fuel oil (?); Aerated water (O_2); NO_3^- (>300 mg/l); PO_4 (>0.3 mg/l); NH_4^+ (>1.0 mg/l)	(1) Removal was fastest for benzene, slower for toluene and slowest for *p*-xylene. (2) Oxygen suspected to be electron acceptor initiating the transformation.

7.1.4. Primary Repository of Expertise

Application of the technology requires a broad range of expertise including microbiology, ground-water hydrology, geochemistry and engineering. Such expertise is available only at a few U.S. governmental laboratories (Environmental Protection Agency and Geological Survey) and U.S. and European universities. Anaerobic microbiology of pollutant transformation is being studied by a growing number of academic and governmental research laboratories.

7.2.   CONTAMINATION THAT IS SUBJECT TO TREATMENT

7.2.1. Chemical Nature

Denitrification has been shown to be effective only for monocyclic and polycyclic aromatic hydrocarbons. Aliphatic hydrocarbon compounds appear to be nondegradable in anaerobic systems.

7.2.2. Range of Concentration

The upper and lower concentration limits for which the technology applies have not yet been defined with certainty. Berry-Spark et al. (1988) were not able to identify a lower limit of BTX removal in a field study where BTX compounds were artificially injected. Most laboratory studies using single substrates or substrate mixtures were conducted with compound concentrations in the mg/l range.

7.3.   REQUIREMENTS FOR SITE CHARACTERIZATION AND IMPLEMENTATION OF THE TECHNOLOGY

For designing and implementing the bioremediation plan, the site has to be characterized with respect to physical (hydrologic), chemical, and biological characteristics. Most of these characteristics are generic to all bioremediation applications.

Physical Characteristics:

1)   Spatial distribution of contamination, especially the distribution of nonaqueous phase material, origin of waste materials, and plume geometry.

2)   Aquifer hydrology, direction and velocity of ground-water flow, heterogeneities, and impermeable zones.

3)   Temperature.

Chemical Characteristics:

1)   Composition of contamination, including nonhydrocarbon contaminants that could interfere with the process.

2)   Ground-water quality, especially redox status, electron acceptors, pH, and degradation products.

3)   Sorption properties of the sedimentary solids for establishing bioavailability.

Biological Characteristics:

1)   Presence of viable microorganisms, especially hydrocarbon degraders, in the contaminated and uncontaminated zones.

2)   Biodegradation potential, degradation rates.

3)   Limiting factors, including limiting nutrients.

7.4.   FAVORABLE SITE CHARACTERISTICS

Site characteristics that are generally favorable for bioremediation include shallow and permeable aquifers, which can readily be supplied with nutrients without excessive pumping costs. However, these characteristics are also favorable for other remediation technologies such as soil gas extraction, or excavation. Anaerobic bioremediation technology, passive or active, should be of greater advantage at sites that are not amenable to treatment with conventional technologies, such as deep or otherwise inaccessible sites.

7.5.   UNFAVORABLE SITE CHARACTERISTICS

7.5.1. Chemical and Physical Nature of the Contamination

1)   NAPLs may act as a long-term source for contaminants and may be toxic for microorganisms.

2)   Mixed wastes may inhibit adaptation of native organisms.

3)   Inhibition by toxic metals. Bollag and Barabasz (1979) and Werner (1985) report that heavy metals such as cadmium, copper, lead and zinc impair denitrifying activity.

4)   Water quality characteristics, i.e. pH or salt content, are incompatible with bacterial physiology.

7.5.2. Site Hydrogeology and Source Characteristics

1)   Aquifer heterogeneity and clay lenses may make contaminated pockets inaccessible for treatment.

2)   Source boundaries are unknown.

3)   Dissolution of minerals may lead to secondary water quality problems.

7.5.3. Infrastructure and Institutional Issues

1)   Efficiency of technology is unproven.

2)   Potentially harmful by-products or end-products may be formed.

7.6.   OPTIMAL SITE CONDITIONS

Optimal growth condition for anaerobic bacteria growing on aromatic hydrocarbons have yet to be determined. Of specific interest are biomass, biomass diversity, biomass yields, temperature, bioavailability, nutrients, formation of inhibiting intermediates, end- or by-products (Texas Research Institute, 1982) and factors specifically related to ground-water conditions such as surface attachment and mobility. Sims et al. (1992) and Wilson, J. et al. (1986) list the following issues:

1)   The concentration of required nutrients in the mobile phase.

2)   The advective flow in the mobile phases or the steepness of concentration gradients within the phases.

3)   Opportunity for colonization of microbially active organisms capable of contaminant degradation.

4)   Co-occurrence of waste materials that may inhibit biotransformation.

7.7.   PROBLEMS ENCOUNTERED WITH THE TECHNOLOGY

Data to evaluate problems with the technology are very sparse. Only data on denitrifying systems have been reported in the literature and only for very few sites. The main concern of these studies has been the efficacy of the process and not potential problems of the technology. Because failed attempts to remediate a site based on denitrification have not been documented, it is difficult to state whether application of the technology is limited by some inherent problems, such as the formation of intermediates and products of health concern. Werner (1991) states without elaboration that the technology "cannot be transferred to the general practice of bioremediation."

Some specific limitations have been reported by Hutchins et al. (1989; 1991b):

-       Only aromatics are subject to treatment.

-       Higher molecular weight compounds, which were sorbed more strongly, were not degraded.

-       Leaching of the more soluble compounds by the nutrient feed water appeared to be a major removal mechanism and may pose secondary disposal problems.

7.8.   PROPERTIES OF SITE AND CONTAMINANTS DETERMINING THE COST OF REMEDIATION

Since supply of nutrient solutions is expected in all systems, hardware installation costs of anaerobic systems are expected to be the same as for aerobic systems. Pumping costs should be lower, however, due to higher electron acceptor concentrations and, consequently, lower nutrient solution volumes. This advantage could be diminished if cleanup times caused by slow growth rates of anaerobes are very long. In any case, source characterization, contaminant distribution and determination of the hydrologic conditions, i.e., site heterogeneity and co-occurrence of mixed wastes, are major cost factors and are the same as those for conventional pump-and-treat and aerobic bioremediation. Thus, anaerobic processes are most likely of greatest advantage in passive remediation schemes where the electron acceptors naturally present can be utilized and the costs of external nutrient supply can be avoided.

7.9. PREVIOUS EXPERIENCE WITH COST OF IMPLEMENTING THE TECHNOLOGY

Hutchins et al. (1989, 1991b) evaluated treatment costs of the Traverse City bioremediation project, which used nitrate as the electron acceptor. They calculated unit costs for the remediation with respect to (1) volume of JP-4 fuel removed, (2) volume of contaminated aquifer material treated, and (3) total aquifer material treated and considered costs for construction, labor, chemicals, and electrical service. The unit costs for the remediation were $22 per liter JP-4, $200 per m^3 aquifer material contaminated with JP-4, and $17 per m^3 of aquifer material down to the confining aquifer. Of course, to assess the viability of the technology, the costs of this technology must be compared with other technologies including conventional and passive treatment technologies.

7.10. FACTORS DETERMINING REGULATORY ACCEPTANCE OF THE TECHNOLOGY

Regulatory acceptance of the technology is limited by:

1) Inadequate understanding of the underlying science, including the chemical, biological and hydrologic factors that guarantee the success of the technology.

2) Lack of successfully completed and documented field demonstration projects.

3) Unpredictable transformation rates leading to uncertain cleanup times and poor process control.

4) Uncertain environmental impact, potential formation of harmful by- and end products.

5) Lack of treatment objectives commensurate with the capabilities of the treatment technologies.

6) Nitrate is regulated with respect to the National Drinking Water Standard. The maximum contaminant level for nitrate is 10.0 mg/l measured as nitrogen.

These potentially unfavorable factors may be compensated for by reduced environmental impacts associated with conventional remediation technologies (e.g., excavation and off-site disposal).

7.11. PRIMARY KNOWLEDGE GAPS AND RESEARCH OPPORTUNITIES

Because anaerobic biotransformation of aromatic hydrocarbon compounds has been reported only recently, details of the process such as characteristics of the organisms and microbial communities, and factors which effect rates and adaptation times, are not yet sufficiently understood. As the published reports are getting more and more detailed, increasingly specific questions can be asked. The questions listed below have been excerpted from the cited reports.

7.11.1. Degradation Under Ideal Conditions:

1)   In what **habitats** do we find microorganisms capable of transforming aromatic hydrocarbon compounds, specifically,
   - to what extent is pre-exposure of the inoculum to hydrocarbons and exact reproduction of the site-geochemical conditions a prerequisite for aromatic hydrocarbon transformation capability?
   - what are the geochemical characteristics of these habitats?

2)   What are the **nutritional requirements** of the microorganisms that are capable of anaerobically degrading aromatic hydrocarbons?

3)   What are typical **lag-times**, specifically

   - how do lag times depend on factors such as substrate structure, co-occurrence and concentration of other more easily degradable substrates and environmental factors?

4)   What are the **degradation rates** under ideal conditions and how are they influenced by environmental factors such as the presence of oxidants, temperature, pH?

5)   What are the **pathways** by which aromatic hydrocarbon compounds are degraded and what intermediates and dead-end products are formed?

6)   What products and intermediates are toxic?

7.11.2. Degradation Under Ground-water and Soil Conditions:

1)   What degradation rates can be expected in porous, sorbing media?

2)   How do redox-active solids such as iron(III) oxides and iron sulfides influence the process?

3)   What abiotic processes are linked to the biodegradation?

4)   How can nutrient conditions and environmental factors be improved?

5)   Are humic and fulvic acids serving as electron acceptors?

7.11.3. Degradation at Sites

1)   How can favorable growth conditions be stimulated under site conditions?

2)   What is the minimum aquifer permeability?

3)   What is the effect of aquifer heterogeneity?

4)   What is the effect of nonaqueous phase liquid hydrocarbons?

7.11.4. Degradation at the Hydrocarbon/Water Interface and Within the Nonaqueous Phase

1) Are organisms growing at the oil/water interface and/or within the nonaqueous phase?

2) What factors are limiting the growth of these bacteria?

3) What is the significance of these organisms in bioremediation schemes?

7.11.5. Methods for Monitoring Performance of Bioremediation Process

1) What in-situ methods are suitable for monitoring the process efficiency?

7.11.6. Design of Optimal Nutrient and Electron Acceptor Systems

1) What are the best nutrient feed systems considering local hydrogeological and geochemical factors?

ACKNOWLEDGEMENT

I thank Harry A. Ball and Harry R. Beller for helpful dicussions of the manuscript.

REFERENCES

Altenschmidt, U., and G. Fuchs. 1991. Anaerobic degradation of toluene in denitrifying *Pseudomonas* sp.: Indication for toluene methylhydroxylation and benzoyl-CoA as central aromatic intermediate. *Arch. Microbiol.* 156:152-158.

Atlas, R.M. 1981. Microbial degradation of petroleum hydrocarbons: An environmental perspective. *Microbiol. Rev.* 45(1):180-209.

Baedecker M.J., and I.M. Cozzarelli. 1991. *Geochemical Modeling of Organic Degradation Reactions in an Aquifer Contaminated with Crude Oil.* U.S. Geological Survey Water-Resources Investigations Report 91-4034. Reston, Virginia. pp. 627-632.

Ball, H.A., M. Reinhard, and P.L. McCarty. 1991. Biotransformation of monoaromatic hydrocarbons under anoxic conditions. In: *In Situ Bioreclamation, Applications and Investigations for Hydrocarbon and Contaminated Site Remediation.* Eds., R.E. Hinchee and R.F. Olfenbuttel. Butterworth-Heinemann. Boston, Massachusetts. pp. 458-463.

Beller, H. R., D. Grbic-Galic, and M. Reinhard. 1992. Microbial degradation of toluene under sulfate-reducing conditions and the influence of iron on the process. *Appl. Environ. Microbiol.* 58:(3)786-793.

Bollag, J.M., and W. Barabasz. 1979. Effects of heavy metals on the denitrification process. *J. Environ. Quality.* 8(2):196-201.

Berry-Spark, K.L. J.F. Barker, K.T. MacQuarrie, D. Major, C.I. Mayfield, and E.E. Sudicky. 1988. *The Behavior of Soluble Petroleum Product Derived Hydrocarbons in Groundwater, Phase III.* PACE Report No. 88-2. Petroleum Association for Conservation of the Canadian Environment. Ottawa, Ontario. Canada.

Dolfing J., J. Zeyer, P. Blinder-Eicher, and R.P Schwarzenbach. 1990. Isolation and characterization of a bacterium that mineralizes toluene in the absence of molecular oxygen. *Arch. Microbiol.* 154:336-341.

Edwards, E., L.E. Wills, D. Grbic-Galic, and M. Reinhard. 1991. Anaerobic degradation of toluene and xylene--evidence for sulfate as the terminal electron acceptor. In: *In Situ Bioreclamation, Applications and Investigations for Hydrocarbon and Contaminated Site Remediation.* Eds., R.E. Hinchee and R.F. Olfenbuttel. Butterworth-Heinemann. Boston, Massachusetts. pp.463-471.

Edwards, E.A., L.E. Wills, M. Reinhard, and D. Grbic-Galic. 1992. Anaerobic degradation of toluene and xylene by aquifer microorganisms under sulfate-reducing conditions. *Appl. Environ. Microbiol.* 58:(3)794-800.

Edwards, E.A., and D. Grbic-Galic. 1992. Complete mineralization of benzene by aquifer microorganisms under strictly anaerobic conditions. *Appl. Environ. Microbiol.* 58(8):2663-2666.

Evans, P.J., D.T. Mang, and L.Y. Young. 1991a. Degradation of toluene and *m*-xylene and transformation of *o*-xylene by denitrifying enrichment cultures. *Appl. Environ. Microbiol.* 57:(2)450-454.

Evans, P.J., D.T. Mang, K.S. Kim, and L.Y. Young. 1991b. Anaerobic degradation of toluene by a denitrifying bacterium. *Appl. Environ. Microbiol.* 57:(4)1139-1145.

Evans, P.J., W. Ling, B. Goldschmidt, E.R. Ritter, and L.Y. Young. 1992. Metabolites formed during anaerobic transformation of toluene and *o*-xylene and their relationship to the initial steps of toluene mineralization. *Appl. Environ. Microbiol.* 58(2):496-501.

Flyvberg, J., E. Arvin, B.K. Jensen, and S.K. Olson. 1991. Bioremediation of oil- and creosote-related aromatic compounds under nitrate-reducing conditions. In: *In Situ Bioreclamation, Applications and Investigations for Hydrocarbon and Contaminated Site Remediation.* Eds., R.E. Hinchee and R.F. Olfenbuttel. Butterworth-Heinemann. Boston, Massachusetts. pp. 471-479.

Godsy, E.M., and D. Grbic-Galic. 1989. Biodegradation pathways for benzothiophene in methanogenic microcosms. In: *U.S. Geological Survey Toxic Substances Hydrology Program: Proceedings of the Technical Meeting.* U.S. Geological Survey Water-Resources Investigations Report 88-4220. Eds., G.E. Mallard and S.E. Ragone. Phoenix, Arizona. September 26-30, 1988. pp. 559-564.

Grbic-Galic, D. 1989. Microbial degradation of homocyclic and heterocyclic aromatic hydrocarbons under anaerobic conditions. *Dev. Ind. Microbiol.* 30:237-253.

Grbic-Galic, D. 1990. Methanogenic transformation of aromatic hydrocarbons and phenols in groundwater aquifer. *Geomicrobiol. J.* 8:167-200.

Grbic-Galic, D., and T.M. Vogel. 1987. Transformation of toluene and benzene by mixed methanogenic cultures. *Appl. Environ. Microbiol.* 53(2):254-260.

Haag, F., M. Reinhard, and P.L. McCarty. 1991. Degradation of toluene and *p*-xylene in an anaerobic microcosms: Evidence for sulfate as a terminal electron acceptor. *Environ. Toxicol. Chem.* 10:1379-1389.

Hutchins S.R., and J.T. Wilson. 1991. Laboratory and field studies on BTEX biodegradation in a fuel-contaminated aquifer under denitrifying conditions. In: *In Situ Bioreclamation, Applications and Investigations for Hydrocarbon and Contaminated Site Remediation.* Eds., R.E. Hinchee and R.F. Olfenbuttel. Butterworth-Heinemann. Boston, Massachusetts. pp. 157-172.

Hutchins, S.R., G.W. Sewell, D.A. Sewell, D.A. Kovacs, and G.A. Smith. 1991a. Biodegradation of aromatic hydrocarbons by aquifer microorganisms under denitrifying conditions. *Environ. Sci. Technol.* 25(1):68-76.

Hutchins, S.R., W.C. Downs, G.B. Smith, J.T. Wilson, D.J. Hendrix, D.D. Fine, D.A. Kovacs, R.H. Douglass, and F.A. Blaha. 1991b. *Nitrate for Biorestoration of an Aquifer Contaminated with Jet Fuel.* U.S. Environmental Protection Agency. Robert S. Kerr Environmental Research Laboratory. Ada, Oklahoma. EPA/600/2-91/009. April, 1991.

Hutchins, S.R., W.C. Downs, D.H. Kampbell, J.T. Wilson, D.A. Kovacs, R.H. Douglass, and D.J. Hendrix. 1989. Pilot project on biorestoration of fuel-contaminated aquifer using nitrate: Part II - Laboratory microcosm studies and field performance. In: *Proceedings of the Petroleum Hydrocarbons and Organic Chemicals in Ground Water: Prevention, Detection, and Restoration*. National Water Well Association. Dublin, Ohio. pp. 589-604.

Kuhn, E.P., J. Zeyer, P.Eicher, and R.P. Schwarzenbach. 1988. Anaerobic degradation of alkylated benzenes in denitrifying laboratory aquifer columns. *Appl. Environ. Microbiol.* 54(2):490-496.

Lee, M.D., J.M. Thomas, R.C. Borden, P.B. Bedient, and J.T. Wilson. 1988. Biorestoration of aquifers contaminated with organic compounds. *CRC Critical Reviews in Environmental Control*. 18:29-89.

Lovley, D.R., M.J. Baedecker, D.J. Lonergan, I.M. Cozzarelli, E.J.P. Phillips, and D.I. Siegel. 1989. Oxidation of aromatic contaminants coupled to microbial iron reduction. *Nature*. 339(6222): 297-300.

Lovley, D.R., and D.J. Lonergan. 1990. Anaerobic oxidation of toluene, phenol and *p*-cresol by the dissimilatory iron-reducing organisms, GS-15. *Appl. Environ. Microbiol*. 56(6):1858-1864.

Major, D.W., C.I. Barker, and J.F. Barker. 1988. Biotransformation of benzene by denitrification in aquifer sand. *Ground Water*. 26(1):8-14.

Mihelcic, J.R., and R.G. Luthy. 1988. Degradation of polycyclic aromatic hydrocarbon compounds under various redox conditions in soil-water systems. *Appl. Environ. Microbiol*. 54(5):1182-1187.

Reinhard, M., N.L. Goodman, and J.F. Barker. 1984. Occurrence and distribution of organic chemicals in two landfill leachate plumes. *Environ. Sci. Technol*. 18(12):953-961.

Reinhard, M., L.E. Wills, H.A. Ball, T. Harmon, D.W. Phipps, H.F. Ridgeway, and M.P. Eisman. 1991. A field experiment for the anaerobic biotransformation of aromatic hydrocarbon compounds at Seal Beach, California. In: *In Situ Bioreclamation, Applications and Investigations for Hydrocarbon and Contaminated Site Remediation*. Eds., R.E. Hinchee and R. Olfenbuttel. Butterworth-Heinemann. Boston, Massachusetts. pp. 487-596.

Sims, J.L., J.M. Suflita, and H.H. Russell. 1992. *In Situ Bioremediation of Contaminated Ground Water*. Ground Water Issue. EPA/540/S-92/003.

Staps, S.J.J.M. 1990. *International Evaluation of In Situ Biorestoration of Contaminated Soil and Groundwater*. EPA 540/2-90/012. September 1990.

Texas Research Institute. 1982. *Enhancing the Microbial Degradation of Underground Gasoline by Increasing Available Oxygen*. API Publication 4428, American Petroleum Institute. Washington, DC.

Werner, P. 1985. A new way for the decontamination of aquifers by biodegradation. *Water Supply*. 3:41-47.

Werner, P.   1991.   German experiences in the biodegradation of creosote and gaswork-specific substances.   In: *In Situ Bioreclamation, Applications and Investigations for Hydrocarbon and Contaminated Site Remediation*.   Eds., R.E. Hinchee and R. Olfenbuttel.   Butterworth-Heinemann.   Boston, Massachusetts. pp. 496-517.

Wilson, B.H., G.B. Smith, and J.F. Rees.   1986.   Biotransformations of selected alkylbenzenes and halogenated aliphatic hydrocarbons in methanogenic aquifer material: a microcosm study.   *Environ. Sci. Technol.*   20(10):997-1002.

Wilson, B.H., B. Bledsoe, and D.H. Kampbell.   1987. Biological processes occurring at an aviation gasoline spill site.   In: *Chemical Quality and the Hydrologic Cycle*.   Eds., R.C. Averett and D.M. McKnight.   Lewis Publishers.   Chelsea, Michigan.   pp. 125-137.

Wilson, J.T., L.E. Leach, M. Henson, and J.N. Jones.   1986.   In situ biorestoration as a ground water remediation technique.   *Ground Water Monitoring Review.*   6(4):56-64.

SECTION 8

BIOREMEDIATION OF CHLORINATED SOLVENTS USING ALTERNATE ELECTRON ACCEPTORS

Edward J. Bouwer
Department of Geography and Environmental Engineering
The Johns Hopkins University
Baltimore, Maryland 21218
Telephone: (410)516-7437
Fax: (410)516-8996

8.1. INTRODUCTION

The contamination of ground water and soils with chlorinated solvents, such as trichloroethene (TCE), tetrachloroethene (PCE), carbon tetrachloride (CT), 1,1,1-trichloroethane (1,1,1-TCA), and chloroform (CF), is widespread (Pye et al., 1983). Their extensive production and use makes these compounds among the most prevalent contaminants in ground water at waste disposal sites. Over 270 million metric tons of the fifty most widely used chemicals were produced in 1988 (Chem. Eng. News, 1989). Synthetic organic compounds including chlorinated solvents accounted for approximately one-third of the chemical production. Many of these chlorinated compounds are known or potential threats to public health and the environment, so there is an urgent need to understand their fate in the environment and develop effective control methods. Flushing the subsurface with water so that chlorinated solvents can dissolve and be pumped to the surface for aboveground treatment is used most frequently for remediation. Because the subsurface is geologically complex and chlorinated solvents tend to sorb to soils, they are not readily leached from the soil, and such pump-and-treat systems are generally inefficient and slow. Furthermore, most aboveground treatment technologies involve physical/chemical processes (e.g., air stripping and carbon adsorption) that simply sequester the contaminants or transfer them to another environmental medium.

Biological processes offer the prospect of converting organic contaminants to harmless products. This cleanup approach, termed bioremediation, stimulates the growth of indigenous or introduced microorganisms in regions of subsurface contamination and, thus, provides direct contact between microorganisms and the dissolved and sorbed contaminants for biotransformation. The process typically entails perfusion of nutrients and one or more electron donors or electron acceptors through the contaminated soil. Certain chlorinated solvents are biotransformed by methanotrophic bacteria under aerobic conditions, and such aerobic bioremediation has been demonstrated to be successful on a small scale in the field (Section 5). However, the aerobic methanotrophic bacteria cometabolize the chlorinated solvents while using methane as a primary substrate. The limited solubility of methane and oxygen in water and competition between methane and the chlorinated solvent for the initial enzyme can restrict the growth of microorganisms and biodegradation of contaminants in the vicinity of a spill, thus reducing the success of aerobic bioremediation. Hydrogen peroxide can be used to increase the oxidant capacity, but this also has disadvantages, including its toxicity to microorganisms and its reactivity with aquifer materials.

Anaerobic bioremediation where electron acceptors other than oxygen are used is potentially advantageous for overcoming the difficulty in supplying oxygen for aerobic processes. Nitrate, sulfate, and carbon dioxide are attractive alternatives to oxygen as an electron acceptor because they are very soluble in water, inexpensive, and nontoxic to microorganisms. Their high aqueous solubility and low reactivity relative to oxygen make them easier to distribute throughout a contaminated zone. Fe(III) and Mn(IV) might be present in the mineral phases of aquifer solids and could serve as alternate electron acceptors for iron- and manganese-reducing bacteria, respectively. Exploiting anaerobic microbial processes for bioremediation of chlorinated solvents is in its infancy. Demonstration of this technology in the field is limited; therefore, the use of alternate electron acceptors for bioremediation of chlorinated solvents must be viewed as a developing treatment technology. Establishing the utility of anaerobic bioremediation for chlorinated solvents is an important scientific and engineering challenge.

This section addresses some important issues concerning the transformation of chlorinated solvents in the absence of oxygen that can be applied to the problem of environmental contamination as well as to the development of engineered treatment processes for subsurface cleanup. These include a discussion of metabolism and biotransformation of chlorinated solvents with alternate electron acceptors, approaches for treatment, reaction stoichiometry, biotransformation rates, and limitations.

8.2.   METABOLISM AND ALTERNATE ELECTRON ACCEPTORS

Biotransformations are driven by the ultimate goal of increasing the size and mass of microbial populations. Microorganisms must transform environmentally available nutrients to forms that are useful for incorporation into cells and synthesis of cell polymers. In general, cells utilize reduced forms of nutrients for these synthesis reactions. Reducing nutrients requires energy and a source of electrons. An electron donor is essential for growing cells; energy is made available for cell growth when the electron donor transfers its electrons to a terminal electron acceptor. Following is an example of a biotransformation in which an organic contaminant typified as benzene (C_6H_6) serves as electron donor and is oxidized to innocuous compounds and supports microbial growth:

$$C_6H_6 + 7.5\ O_2 \longrightarrow 6\,CO_2 + 3\,H_2O \tag{1a}$$

$$C_6H_6 + 1.5\ HCO_3^- + 1.5\ NH_4^+ \longrightarrow 1.5\ C_5H_7O_2N + 1.5\ H_2O \tag{1b}$$

In Reaction 1a, the transfer of electrons between benzene (electron donor) and O_2 (electron acceptor) provides energy for synthesis of cellular material $(C_5H_7O_2N)$ from the benzene carbon (Reaction 1b). By this process, a portion of an organic contaminant serves as a primary energy source that is converted to end products, and a portion of the contaminant carbon is synthesized into biomass.

The terminal electron acceptor used during metabolism is important for establishing the redox conditions and the chemical speciation in the vicinity of the cell. Common terminal electron acceptors include oxygen under aerobic conditions, and nitrate, Mn(IV), Fe(III), sulfate, and carbon dioxide under anaerobic conditions. Microorganisms preferentially utilize electron acceptors that provide the maximum free energy during respiration. Of the common electron acceptors used by microorganisms, oxygen has the highest redox potential and provides the most free energy to microorganisms during electron transfer (Figure 8.1). The redox potentials of nitrate, Mn(IV), Fe(III), sulfate, and carbon dioxide are lower (Figure 8.1). Consequently, they yield less energy during substrate oxidation and electron transfer according to the order listed in Figure 8.1. These latter compounds comprise the alternate electron acceptors available for development of anaerobic bioremediation technologies. The importance of microbial reactions involving Mn(IV) and Fe(III) to organic contaminant biotransformations is unknown. Therefore, this section will focus on microbial systems involving nitrate (denitrification), sulfate (sulfate reduction), and carbon dioxide (methanogenesis) as electron acceptors.

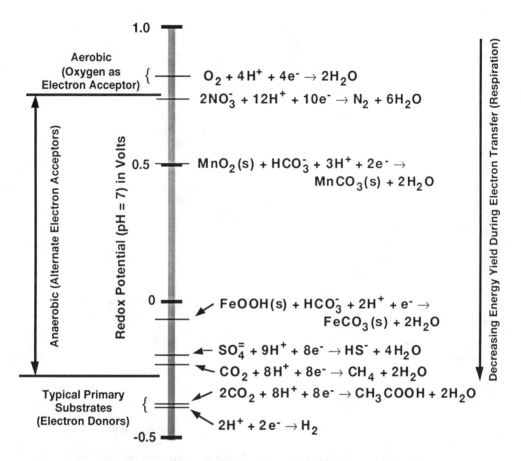

Figure 8.1.   Important electron donors and acceptors in biotransformation processes. Redox potentials were obtained from Stumm and Morgan (1981).

8.3. BIOTRANSFORMATION OF CHLORINATED SOLVENTS IN THE PRESENCE OF ALTERNATE ELECTRON ACCEPTORS

The redox environment is an important factor affecting microbial respiration and biotransformation of organic contaminants. Some compounds are only transformed under aerobic conditions; others require strongly reducing conditions; and still others are transformed in both aerobic and anaerobic environments. Reviews of aerobic and anaerobic biotransformations of petroleum hydrocarbons and aerobic biotransformation of chlorinated solvents are presented in other sections within this volume. Examples of how metabolism with alternate electron acceptors influences the biotransformations of some chlorinated solvents of concern are described in this section. This knowledge coupled with the spatial distribution of electron acceptors and other redox species within a region of subsurface contamination is important for identifying zones conducive to biotransformation of a particular chlorinated solvent. The coupling of redox conditions and chlorinated solvent biotransformation is also important in establishing how to chemically manipulate the medium to achieve a desired biotransformation.

In the absence of molecular oxygen, microbial reduction reactions involving organic contaminants increase in significance as environmental conditions become more reducing. Nearly 25 years ago, Castro and Belser (1968) found that the soil fumigants ethylene dibromide (1,2-dibromoethane), 1,2-dibromo-3-chloropropane, and 2,3-dibromobutane were transformed in soil slurries via a reductive dehalogenation reaction. In reductive dehalogenation, the halogenated compound becomes an electron acceptor; and in this process, a halogen is removed and is replaced with a hydrogen atom. About a decade later, investigations were initiated to evaluate the fate of chlorinated solvents (mainly chloromethanes and chloroethenes) in anaerobic environments. The results of investigations with microcosms and enrichments from environmental samples under anaerobic conditions are summarized in Table 8.1. Anaerobic biotransformation of chlorinated solvents has been observed in field studies (Roberts et al., 1982), in continuous-flow fixed-film reactors (Bouwer and McCarty, 1983b; Vogel and McCarty, 1985, 1987; Bouwer and Wright, 1988), and in soil (Kloepfer et al., 1985), sediment (Barrio-Lage et al., 1986), and aquifer microcosms (Wilson, B. et al., 1986) under conditions of denitrification, sulfate reduction, or methanogenesis. The initial step in the anaerobic biotransformation was generally reductive dechlorination. For example, CF was produced from CT, and 1,1-dichloroethane (1,1-DCA) was produced from 1,1,1-TCA (Table 8.1).

The transformations of PCE and TCE have been studied most intensely. General agreement exists that transformation of these two compounds under anaerobic conditions proceeds by sequential reductive dechlorination to dichloroethene (DCE) and vinyl chloride (VC); and in some instances, there is total dechlorination to ethene or ethane. Of the three possible DCE isomers, 1,1-DCE is the least significant intermediate; several studies have reported that *cis*-1,2-DCE predominates over *trans*-1,2-DCE (Barrio-Lage et al., 1986; Parsons et al., 1984; Parsons and Lage, 1985). CT, CF, 1,2-DCA, 1,1,1-TCA, and PCE were partially converted to carbon dioxide during the anaerobic biotransformations. Reductive dechlorination of 1,1,1-TCA and PCE occurred first prior to mineralization to carbon dioxide. Most of the experiments were conducted under methanogenic conditions. Several of the chlorinated compounds were also transformed by similar pathways under conditions of denitrification and sulfate reduction (Table 8.1).

Several studies provide evidence for anaerobic transformation of chlorinated solvents by pure cultures of bacteria (Table 8.2). The bacteria involved ranged from strict anaerobic microorganisms, such as methanogens, sulfate-reducers, and clostridia to facultative anaerobes such as *Escherichia coli* or *Pseudomonas putida*. Reductive dechlorination was the predominant reaction pathway. Consequently, the chlorinated solvent biotransformation studies with environmental samples (mixed microbial cultures) and pure bacterial cultures indicate that a broad variety of bacteria possess the enzymatic capability to reductively dechlorinate the compounds. An electron donor, such as low molecular weight organic compounds (lactate, acetate, methanol, glucose, etc.) or H_2, must be available to provide reducing equivalents for reductive dechlorination. Toluene was recently found to be a suitable electron donor for the reductive dechlorination of PCE to DCE in anaerobic aquifer microcosms (Sewell and Gibson, 1991).

TABLE 8.1. ANAEROBIC TRANSFORMATION OF SELECTED CHLORINATED SOLVENTS IN MICROCOSMS AND ENRICHMENT CULTURES UNDER DIFFERENT REDOX CONDITIONS

Chlorinated solvent[a]	Redox condition[b]	Transfor- mation[c]	Intermediate[d]	End product	System	Refer- ence(s)[e]
CT	dn	+	CF	n.d.	biofilm reactor	d,e
	sr	+	CF	n.d.	biofilm reactor	e
	me	+	CF	CO_2	biofilm reactor/aquifer material	c,e,n
CF	dn	--	--	--	biofilm reactor	d,e
	sr	--	--	--	biofilm reactor	e
	me	+	n.d.	CO_2	biofilm reactor	c,e
1,2-DCA	me	+	n.d.	CO_2	biofilm reactor	c
1,1,1-TCA	dn	--	--	--	biofilm reactor	d,e,j
	sr	+	1,1-DCA	CA	biofilm reactor/aquifer material	e,j
	me	+	1,1-DCA	CO_2	biofilm reactor/aquifer material	c,e,j,r
1,1,2,2-TeCA	me	+	--	1,1,2-TCA	biofilm reactor	c
HCA	ae	+	--	PCE	aquifer material	f
	dn	+	n.d.	n.d.	biofilm reactor	e
	sr	+	n.d.	n.d.	biofilm reactor	e
	me	+	n.d.	n.d.	biofilm reactor	e
1,1-DCE	me	+	VC	n.d.	sediment/aquifer material	b,s
cis-1,2-DCE	me	+	VC	n.d.	sediment/aquifer material	b,s
trans-1,2-DCE	me	+	CA + VC	n.d.	sediment/aquifer material	b,s
TCE	me	+	cis + trans-1,2-DCE	n.d.	aquifer material	m,n
			1,2-DCE	n.d.	aquifer material	k,s
PCE	sr	+	TCE	cis-1,2-DCE	sewage sludge	a
	me	+	TCE	CO_2	biofilm reactor	q
				ethene	sewage sludge	g,h
				cis + trans-1,2-DCE	aquifer material	p
				cis-1,2-DCE	aquifer material/sewage sludge	i,o
				n.d.	aquifer material	c,m,n,l

a Abbreviations stand for: CT = carbon tetrachloride; CF = chloroform; DCA = dichloroethane; TCA = trichloroethane; CA = chloroethane; TeCA = tetrachloroethane; HCA = hexachloroethane; DCE = dichloroethene; TCE = trichloroethene; PCE = tetrachloroethene.

b ae = aerobic; dn = denitrification; sr = sulfate reduction; me = methanogenesis.

c + = transformation observed; -- = no transformation

d n.d. = not determined

e a = Bagley and Gossett, 1990; b = Barrio-Lage et al., 1986; c = Bouwer and McCarty, 1983a; d = Bouwer and McCarty, 1983b; e = Bouwer and Wright, 1988; f = Criddle et al., 1986; g = DiStefano et al., 1991; h = Freedman and Gossett, 1989; i = Kästner, 1991; j = Klecka et al., 1990; k = Kloepfer et al., 1985; l = Parsons and Lage, 1985; m = Parsons et al., 1984; n = Parsons et al., 1985; o = Scholz-Muramatsu et al., 1990; p = Sewell and Gibson, 1991; q = Vogel and McCarty, 1985; r = Vogel and McCarty, 1987; s = Wilson, et al., 1986.

TABLE 8.2.   REDUCTIVE DEHALOGENATION REACTIONS CATALYZED BY PURE CULTURES OF BACTERIA

Halogenated compound[a]	Bacteria	Products	Reference(s)[b]
CT	*Methanobacterium thermoautotrophicum*	$CF \rightarrow DCM + CO_2$	c,e
	Methanosarcina barkeri	$CF \rightarrow DCM$	i
	Desulfobacterium autotrophicum	$CF \rightarrow DCM + CO_2$	c,d,e
	Acetobacterium woodii	$CF \rightarrow DCM \rightarrow CM + CO_2$	d,e
	Clostridium thermoaceticum	$CF \rightarrow DCM \rightarrow CM + CO_2$	d
	Clostridium sp.	$CF \rightarrow DCM +$ unidentified	h
	Escherichia coli	CF	b
CF	two *Methansarcina* sp.	$DCM \rightarrow CM$	j
1,2-DCA	several methanogens	ethene	a,d
1,1,1-TCA	*Methanobacterium thermoautotrophicum*	1,1-DCA	c
	Desulfobacterium autotrophicum		c
	Acetobacterium woodii		d
	Clostridium sp.	1,1-DCA + acetate + unidentified	h
BA	several methanogens	ethane	a
1,2-DBA	several methanogens	ethane	a
PCE	several methanogens	TCE	c,f,g
	Desulfomonile tiedjei		g
	Acetobacterium woodii		d
1,2-DBE	several methanogens	acetylene	a

[a] Abbreviations stand for: CT = carbon tetrachloride; CF = chloroform; DCA = dichloroethane; TCA = trichloroethane; BA = bromoethane; DBA = dibromoethane; DBE = dibromoethene; CM = chloromethane; DCM = dichloromethane; TCE = trichloroethene; PCE = tetrachloroethene.

[b] a = Belay and Daniels, 1987; b = Criddle et al., 1990a; c = Egli et al., 1987; d = Egli et al., 1988; e = Egli et al., 1990; f = Fathepure and Boyd, 1988; g = Fathepure et al., 1987; h = Galli and McCarty, 1989; i = Krone et al., 1989; j = Mikesell and Boyd, 1990

Conversion of the chlorinated aliphatic compounds to less chlorinated alkenes and alkanes via reductive dechlorination is of little or no benefit in the context of anaerobic bioremediation.   The intermediates commonly observed, such as *cis*-1,2-DCE, *trans*-1,2-DCE, VC, 1,1-DCA, and CF, also pose a threat to public health. The possible formation of toxic metabolites has been the major impediment to the development of practical anaerobic bioremediation in the field for cleanup of chlorinated solvent contamination. For anaerobic bioremediation to be useful, chlorinated solvents must be biotransformed to nonchlorinated, environmentally acceptable products. Some recent laboratory studies have demonstrated that this is possible and help provide impetus to further develop anaerobic biological processes for bioremediation.

8.3.1. Carbon Tetrachloride Biotransformation

A denitrifying *Pseudomonas* sp. (strain KC) capable of rapid and complete biotransformation of CT was isolated from aquifer material (Criddle et al., 1990a). This bacterium was able to completely convert CT to carbon dioxide and an unidentified water-soluble fraction without simultaneous production of CF. Denitrification was confirmed by consumption of nitrate and acetate as primary substrate and production of protein. This CT biotransformation could potentially be exploited in anaerobic bioremediation of contaminated ground water. The use of denitrifying organisms would be advantageous because nitrate is highly soluble in water and easily added. When reduced iron and cobalt were provided in the growth medium, CT biotransformation to mineralized products was inhibited. Consequently, careful attention must be paid to the trace-metal availability in the engineering of systems that direct CT to nonhazardous end products.

8.3.2. Tetrachloroethene and Trichloroethene Biotransformation

Reductive dechlorination of PCE, via TCE, *cis*-1,2-DCE, and VC to ethene was observed at 20°C in a laboratory-scale fixed-bed column packed with a mixture (3:1) of anaerobic sediment from the river Rhine and anaerobic granular sludge (de Bruin et al., 1992). In the presence of lactate (1 mM) as an electron donor, 9 µM PCE was dechlorinated to ethene. Ethene was further reduced to ethane. Nearly complete conversion (95 - 98%) of PCE occurred in the bioactive anaerobic column with no chlorinated products remaining (< 0.5 µg/l). A novel bacterium, designated "PER-K23," was enriched from the anaerobic column that catalyzed the dechlorination of PCE via TCE to *cis*-1,2-DCE and coupled this reductive dechlorination to growth (Holliger, 1992). H_2/CO_2 or formate were the only electron donors that supported growth with PCE or TCE as electron acceptors. PER-K23 did not grow in the absence of PCE or TCE. PCE or TCE could not be replaced with oxygen, nitrate, nitrite, sulfate, sulfite, thiosulfate, sulfur, fumarate, or carbon dioxide as electron acceptor with H_2 as electron donor. All electrons derived from H_2 or formate consumption could be recovered in dechlorination products and biomass formed. This dependence on a chlorinated hydrocarbon as an electron acceptor is an important step in reducing the chemical requirements (electron donor) and increasing the reaction rate of anaerobic bacterial reductive dehalogenation. The isolation of this novel organism is an important initial step in the development of stable cultures for converting chlorinated solvents to harmless products.

Anaerobic enrichment cultures obtained from wastewater digested sludge which support methanogenesis were capable of completely dechlorinating PCE and TCE via 1,2-DCEs and VC to ethene without significant conversion to CO_2 or CH_4 (Freedman and Gossett, 1989). The rate-limiting step in the transformation sequence appeared to be conversion of VC to ethene. It was necessary to supply an electron donor, such as methanol, hydrogen, formate, acetate, or glucose, to sustain reductive dechlorination of PCE and TCE. Additional studies with these enrichment cultures yielded an anaerobic microbial system capable of dechlorinating PCE as high as 550 µM to 80% ethene and 20% VC within 2 days at 35°C (DiStefano et al., 1991). Methanol was required for this conversion of PCE to ethene at a concentration level approximately twice that needed for complete dechlorination of PCE to ethene. When the incubation was allowed to proceed for as long as 4 days, virtually complete conversion of PCE to ethene resulted, with <1% of the initial 550 µM PCE (250 µg/l) persisting as VC. These findings are encouraging for anaerobic bioremediation of PCE-contaminated sites because a high initial volumetric PCE dechlorination rate was observed at 35°C (275 μM/day) and a relatively large fraction (about one-third) of the supplied electron donor was used for dechlorination. Whether such rates and stoichiometry are possible at lower temperatures typical for the subsurface remain unknown.

8.4.   APPROACHES FOR TREATMENT

The previous section has indicated that under certain environmental conditions with alternate electron acceptors, CT is mineralized to carbon dioxide and PCE is completely dechlorinated to ethene and ethane. These favorable reactions could be exploited in subsurface bioremediation by involving the following steps: (1) characterization of site hydrogeology and contamination, (2) removal of any separate immiscible phase, (3) assessment of biotransformation, (4) system design and operation, and (5) monitoring of system performance (Thomas and Ward, 1989). These steps are developed fully in earlier sections within this volume; only an overview is presented here. Information regarding site geology and hydrology must be defined in order to properly determine the eventual location of the treatment system. Geological considerations should include stratigraphic effects such as horizontal extent of the aquifer and heterogeneity of the soil. Hydrogeological data include porosity, permeability, and ground-water velocity, direction, and recharge/discharge (Freeze and Cherry, 1979). In addition, hydraulic connection between aquifers, potential recharge/discharge areas, and water table fluctuations must be considered. It is important to initially identify the contaminants present and their concentrations, since the microbial systems capable of biotransformation and rates are compound specific.

Bioremediation with alternate electron acceptors will involve the stimulation of microbial growth by perfusion of electron donor, electron acceptor, and nutrients through the formation. The process is most attractive when indigenous bacteria are used, as this avoids the significant problem of injecting and distributing a population of bacteria acclimated to the contaminants. The frequent occurrence of reductive dehalogenation reactions in anaerobic ground waters suggests that microorganisms involved are frequently present in the subsurface. Formations with hydraulic conductivities of 10^{-4} cm/sec or greater are most amenable to bioremediation. Other factors that are considered favorable for applying biotransformation in a cleanup operation are listed in Table 8.3. For comparison, unfavorable conditions for in-situ bioremediation are also included in Table 8.3.

Feasibility studies for the biotransformation of the contaminants are usually conducted in the laboratory using subsurface material collected from the site prior to the design and operation of a full-scale system. These experiments are conducted to establish the presence of microorganisms capable of biotransforming the organic contaminant(s), their nutrient and electron acceptor requirements, and the range of contaminant concentrations that are not completely inhibitory to the microorganisms. Sorption studies should be conducted to determine the extent and rates of partitioning of contaminants onto the aquifer solids. The greater the degree to which the contaminants are sorbed, the more difficult it is for the contaminant to come into contact with microorganisms, and the longer the time for cleanup.

TABLE 8.3. FAVORABLE AND UNFAVORABLE CHEMICAL AND HYDROGEOLOGICAL SITE CONDITIONS FOR IMPLEMENTATION OF IN-SITU BIOREMEDIATION[a]

FAVORABLE FACTORS

CHEMICAL CHARACTERISTICS
  small number of organic contaminants
  nontoxic concentrations
  diverse microbial populations
  suitable electron acceptor condition
  pH 6 to 8

HYDROGEOLOGICAL CHARACTERISTICS
  granular porous media
  high permeability (K > 10^{-4} cm/sec)
  uniform mineralogy
  homogeneous media
  saturated media

UNFAVORABLE FACTORS

CHEMICAL CHARACTERISTICS
  numerous contaminants
  complex mixture of inorganic and organic compounds
  toxic concentrations
  sparse microbial activity
  absence of appropriate electron acceptors
  pH extremes

HYDROGEOLOGICAL CHARACTERISTICS
  fractured rock
  low permeability (K < 10^{-4} cm/sec)
  complex mineralogy
  heterogeneous media
  unsaturated-saturated conditions

[a]Wagner et al., 1986

The use of alternate electron acceptors for control of chlorinated solvents is likely to be accomplished in situ. The advantage of in-situ treatment is that the contaminated water or aquifer material do not have to be pumped or transported. It might be possible to establish anaerobic conditions for the treatment of soil and ground water near the land surface by using infiltration galleys that allow substrates and nutrient laden water to percolate through the soil. When contamination is located at greater depths, the approach to in-situ anaerobic bioremediation is accomplished by infiltrating or injecting electron donor, electron acceptor, and nutrients into the contaminated subsurface to stimulate anaerobic microorganisms in the contaminant plume. Alternatively, the necessary growth factors could be injected downgradient from the contaminant plume to establish an anaerobic biological treatment zone. Here, biotransformation of the chlorinated solvent occurs as the plume percolates through the zone of microbial activity. Chemotactic bacteria move in response to chemical agents. When present, chemotactic bacteria may slowly move upgradient in the direction of increasing organic contaminant concentrations. This will expand the zone of biotransformation and allow contact with sorbed and immiscible compounds. A dynamic system that includes injection and extraction wells and equipment for the addition and mixing of nutrients could be used to better control flow and movement of electron donor, electron acceptor, and nutrients and contaminants. The objective is to stimulate anaerobic microorganisms to transform a portion of the desorbed chlorinated compound with each pass of water laden with growth-supporting chemicals. Since the microorganisms colonize the soil surfaces, chlorinated compounds could be biotransformed as they desorb from the aquifer solids. Biotransformation reduces the solution concentration, thus enhancing the rate of desorption or dissolution of an immiscible phase. Periodic sampling of the soil and ground water is essential for determining the progress of the bioremediation.

8.5.   FIELD EXPERIENCE

Bioremediation of chlorinated solvents using alternate electron acceptors is a developing treatment technology that is mostly being investigated at the laboratory scale. Limited field experience exists on stimulation of anaerobic biotransformation for control of chlorinated solvents. One field study demonstrating this technology was conducted at the Moffett Field Naval Air Station, Mountain View, California (Semprini et al., 1991). This site was used earlier to study in-situ restoration of chlorinated aliphatics by methanotrophic bacteria (Roberts et al., 1990). Reducing conditions were promoted in the field in a 2-m test zone by stimulating a consortium of denitrifying bacteria, and perhaps sulfate-reducing bacteria, through the addition of acetate as primary substrate (25 mg/l). The aquifer contained both nitrate (25 mg/l) and sulfate (700 mg/l). CT was continuously injected at a concentration of 40 µg/l, and between 95 and 97 percent CT biotransformation was observed in the 2-m test zone with stimulated anaerobic growth. CF was an intermediate product and represented 30 to 60 percent of the CT transformed. Other halogenated aliphatics were biotransformed, but at slower rates and lower extents of removal. Removals achieved for Freon-11, Freon-113, and 1,1,1-TCA ranged between 65 to 75 percent, 10 to 30 percent, and 11 to 19 percent, respectively.

A second field demonstration was conducted at a chemical transfer facility in North Toronto (Major et al., 1991). The aquifer at this site was contaminated with organic solvents (methanol, methyl ethyl ketone, vinyl and ethyl acetate, and butyl acrylate) and PCE. Samples of the aquifer material were amended with PCE plus acetate/methanol. Over a 145-day incubation period in the laboratory, PCE was dechlorinated to TCE, then *cis*-1,2-DCE, VC, and in many instances, to ethene. From these results the investigators hypothesized that the presence of methanol in the contaminated site serves as primary substrate for complete dechlorination of PCE by anaerobic microorganisms. In-situ anaerobic bioremediation appears to be occurring at the site without addition of chemicals.

8.6.   SEQUENTIAL ANAEROBIC/AEROBIC TRANSFORMATIONS OF CHLORINATED SOLVENTS

The combination of an anaerobic process followed by an aerobic process has promise for bioremediation of highly chlorinated organic contaminants to innocuous products. Generally, anaerobic microorganisms can reduce the number of chlorines on a chlorinated compound via reductive dechlorination as described in a previous section.

Chlorinated compounds are relatively oxidized by the presence of chlorine substituents. The susceptibility to reduction reactions increases as the number of chlorine substituents increases. Conversely, as the number of chlorine substituents decreases on a given organic compound, reductive dechlorination reactions become less rapid and are less likely to occur. Therefore, it is difficult to achieve complete loss of the chlorine substituents by reductive dechlorination under anaerobic conditions. Mono- and dichlorinated compounds tend to accumulate from the transformation of polychlorinated organic compounds under reducing microbial conditions. Only in specialized cases has complete dechlorination been observed via reductive microbial processes.

However, aerobic microorganisms are capable of transforming some chlorinated compounds, especially those with fewer chlorine substituents. This oxidative process often results in complete mineralization to carbon dioxide and is mediated by three general mechanisms: incorporation of oxygen in the carbon-hydrogen bond, oxidation of a halogen substituent, and oxidation of a carbon-carbon double bond via epoxidation. With fewer chlorine substituents, the more reduced the compound, and the more susceptible it is to oxidation. With removal of chlorines, oxidation becomes more favorable than does reductive dechlorination. Therefore, the combination of anaerobic and aerobic processes has potential utility as a control technology for chlorinated solvent contamination. The approach is to first stimulate anaerobic bacteria followed by creating oxic conditions for methanotrophs. In such a sequence the products of an incomplete anaerobic dechlorination could be oxidized by cometabolic reactions involving methanotrophs.

Anaerobic dechlorination and aerobic biodegradation have not been shown to occur in sequence within the same natural system. However, this combination of anaerobic and aerobic reactions for treatment has been tested in the laboratory. PCE and TCE were transformed to DCE under methanogenic conditions in a 23-liter laboratory aquifer simulator containing contaminated soil and ground water (Dooley-Danna et al., 1989). A recirculation flow of glucose and nutrients was used to maintain methanogenic conditions. Oxygen was then introduced and the oxidation of DCE by methanotrophic bacteria was initiated. The sequential anaerobic/aerobic manipulations resulted in complete biotransformation of the PCE and TCE. Hexachlorobenzene, PCE, and CT were dechlorinated to at least the dichlorinated products in a methanogenic biofilm column reactor fed acetate as the primary substrate (Vogel et al., 1989). All of the reductive dechlorination products in the effluent of the methanogenic biofilm reactor were fed to an aerobic biofilm reactor seeded with settled sewage. The mono- and dichlorinated compounds were effectively utilized by the aerobic biofilm. Although sequential anaerobic/aerobic treatment is a promising alternative to overcome the possible accumulation of partially dechlorinated intermediates under anaerobic conditions, alternating reducing and oxidizing conditions will be difficult to achieve in the field.

8.7. PERFORMANCE

8.7.1. Physical/Chemical Properties

An important factor in the success of subsurface biotransformation is the availability of the contaminant for microbial reactions. Important physical chemical properties influencing contaminant availability include density, water solubility, Henry's constant (H), and n-octanol/water partition coefficient (K_{ow}). Such physical chemical data are summarized in Table 8.4 for some chlorinated solvents commonly encountered in ground-water contamination problems.

TABLE 8.4.   PHYSICAL CHEMICAL PROPERTIES OF CHLORINATED SOLVENTS COMMON TO
               GROUND-WATER CONTAMINATION[a]

Compound	Density, g/ml	Solubility, mg/l	Henry's constant, atm	$log_{10}K_{ow}$
trichloroethylene	1.4	1,100	550	2.29
tetrachloroethylene	1.63	200	1,100	2.88
chloroform	1.49	8,200	170	1.95
1,1-dichloroethylene	1.013	250	1,400	0.73
1,1,1-trichloroethane	1.435	480	860	2.49
vinyl chloride	gas	1,100	35,500	0.60
carbon tetrachloride	1.59	800	1,200	2.64

[a] Data obtained from Verschueren (1983), U.S. EPA (1985), and Lyman et al. (1982).

The n-octanol/water partition coefficient (K_{ow}), which characterizes the hydrophobic nature of the compound, indicates the tendency for the compound to partition (sorb) into soil organic matter. Compounds with low solubility and high K_{ow} tend to sorb strongly to aquifer solids, which retard their movement and decrease their availability for biotransformation. Conversely, contaminants with high water solubility and low K_{ow} are quite mobile and can be transported great distances with ground-water flow. Chlorinated solvents fall into the latter class of compounds. Typical values of $log_{10}K_{ow}$ for chlorinated solvents range between 0.6 and 3.0 (Table 8.4). Chlorinated solvents migrate at rates 10% to nearly 100% of the velocity of ground water (Mackay et al., 1985). On a relative basis, chlorinated solvents sorb less strongly than aromatic hydrocarbons common to petroleum mixtures. This is a favorable property for bioremediation. Also, the aqueous solubilities of chlorinated solvents are high, making them readily available as substrates for microorganisms. However, chlorinated solvent sorption is significant enough that effects of desorption coupled with geologic complexity often make extraction problematic in pump-and-treat remediation (Mackay and Cherry, 1989). Furthermore, the high aqueous solubility can lead to inhibitory concentrations in the vicinity of a spill.

The Henry's constants for chlorinated solvents are high (>100 atm), making volatilization an important loss process in open systems. Volatilization can occur in the vadose zone or during soil excavation but is not significant under saturated flow conditions necessary to achieve anaerobic conditions and utilize alternate electron acceptors.

8.7.2. Concentration Range

Chlorinated solvents are inhibitory to anaerobic microorganisms, which restricts the range of concentrations appropriate for treatment. Belay and Daniels (1987) reported that nearly complete inhibition of pure methanogenic cultures occurred for DCA, DCE, and TCE at exposure concentrations in the range of 50 to 150 mg/l. Partial inhibition (20 - 50%) was observed for exposure concentrations in the range of 10 to 50 mg/l. CT toxicity studies with methanogens conducted by Yang and Speece (1986) showed similar findings. Inhibition of unacclimated cultures was noted at 0.5 mg/l; but with acclimation, 15 mg/l could be tolerated. The dechlorinating culture reported by DiStefano et al. (1991) functioned effectively with PCE at 550 μM (91 mg/l). Consequently, the maximum allowable concentration for treatment depends on the specific chlorinated compound and appears to range between 10 and 100 mg/l. Many of the chlorinated solvent biotransformation studies described in Tables 8.1 and 8.2 were performed with initial concentrations less than 1000 μg/l. Inhibitory effects were not observed at these low concentration levels typical of many contaminated ground waters.

Nearly complete removal of the parent chlorinated compound can occur during anaerobic biotransformation. For many of the studies described in Tables 8.1 and 8.2, residual concentrations of the starting chlorinated compound were in the low μg/l or even <1 μg/l in some cases. However, the repeatedly observed incomplete reductive dechlorination results in accumulation of lesser chlorinated compounds and ineffective treatment. The concentrations of chlorinated intermediates and products remaining after anaerobic biotransformation of chlorinated solvents, and consequently the degree of success, are critically linked to how complete the reductive dechlorination reactions have taken place. Concentrations below typical health based standards of 5 to 10 μg/l were not achieved in the two field trials presented on page 8-10. However, the anaerobic microbial systems showing conversion of PCE to ethene (DiStefano et al., 1991; de Bruin et al., 1992) and CT to CO_2 (Criddle et al., 1990b) are especially encouraging for development of this treatment technology. Residual concentrations below 5 μg/l were observed in these laboratory microcosms. Consequently, it appears that optimized systems for anaerobic biotransformation can meet relevant regulatory endpoints.

8.7.3. Favorable Redox Conditions

There are several possible electron acceptors for anaerobic biotransformations. In many subsurface systems, the redox state is governed by microbial activity. Energetic considerations can be used to obtain insight into which chlorinated compounds are most susceptible to reductive dechlorination in the presence of different electron acceptors being used by the microorganisms. For assessing whether a given compound may in principle undergo a redox reaction like reductive dechlorination, one needs to know the (standard) reduction potentials of the half-reactions involving the compound of interest and its oxidized or reduced transformation product, as well as the reduction potentials of the natural oxidant(s) or reductant(s) present in a given system. The (standard) reduction potentials at pH 7 of some important microbial electron acceptors are given in Table 8.5 together with the reduction potentials of some half-reactions involving chlorinated solvents. The half-reactions are ordered in decreasing redox potential values expressed in volts. A favorable thermodynamic redox reaction is obtained by coupling any given reduced species as electron donor with an electron acceptor of higher redox potential (listed above the electron donor half-reaction listed in Table 8.5). For example, acetate as electron donor can be coupled with all of the other half-reactions listed above it to yield a thermodynamically favorable reaction. The conversion of CT to CF, PCE to TCE, 1,1,1-TCA to 1,1-DCA, CF to methylene chloride, TCE to *trans*-1,2-DCE, and *trans*-1,2-DCE to VC appear to be energetically favorable under sulfate-reducing (sulfate as electron acceptor) and methanogenic (CO_2 as electron acceptor) conditions. Reduction of ethylene dibromide to ethene and hexachloroethane to PCE even appears thermodynamically possible under aerobic respiration and denitrification. Several of the reductive dechlorination reactions appear to be thermodynamically possible with Fe(III) and Mn(IV) as electron acceptors.

TABLE 8.5. STANDARD REDUCTION POTENTIALS AT 25°C AND pH 7 FOR SOME REDOX COUPLES THAT ARE IMPORTANT ELECTRON ACCEPTORS IN MICROBIAL RESPIRATION AND FOR SOME HALF-REACTIONS INVOLVING CHLORINATED SOLVENTS

HALF REACTION[a]		$E°$
Oxidized species	*Reduced species*	*(volts)[b]*
$Cl_3C\text{-}CCl_3 + 2e^-$	$= Cl_2C{=}CCl_2 + 2Cl^-$	+1.13
$\mathbf{O_2 + 4H^+ + 4e^-}$	$\mathbf{= 2H_2O}$	+0.82
$\mathbf{2NO_3^- + 12H^+ + 10e^-}$	$\mathbf{= N_2 + 6H_2O}$	+0.74
$CCl_4 + H^+ + 2e^-$	$= CHCl_3 + Cl^-$	+0.67
$Cl_2C{=}CCl_2 + H^+ + 2e^-$	$= HClC{=}CCl_2 + Cl^-$	+0.58
$Cl_3C\text{-}CH_3 + H^+ + 2e^-$	$= HCl_2C\text{-}CH_3 + Cl^-$	+0.57
$CHCl_3 + H^+ + 2e^-$	$= CH_2Cl_2 + Cl^-$	+0.56
$HClC{=}CCl_2 + H^+ + 2e^-$	$= t\text{-}HClC{=}ClH + Cl^-$	+0.54
$\mathbf{MnO_2(s) + HCO_3^- + 3H^+ + 2e^-}$	$\mathbf{= MnCO_3(s) + 2H_2O}$	+0.52
$t\text{-}HClC{=}ClH + H^+ + 2e^-$	$= H_2C{=}CHCl + Cl^-$	+0.37
$\mathbf{FeOOH(s) + HCO_3^- + 2H^+ + e^-}$	$\mathbf{= FeCO_3(s) + 2H_2O}$	-0.05
$\mathbf{SO_4^= + 9H^+ + 8e^-}$	$\mathbf{= HS^- + 4H_2O}$	-0.22
$\mathbf{CO_2 + 8H^+ + 8e^-}$	$\mathbf{= CH_4 + 2H_2O}$	-0.24
$2CO_2 + 8H^+ + 8e^-$	$= CH_3COOH + 2H_2O$	-0.40
$2H^+ + 2e^-$	$= H_2$	-0.41

[a] Half-reactions in bold are common electron acceptors in microbial respiration.

[b] Data from Stumm and Morgan (1981) and Thauer et al. (1977). Values are for aqueous solution with pH = 7, $[HCO_3^-] = 0.001$ M, and $[Cl^-] = 0.001$ M.

8.8. BIOTRANSFORMATION STOICHIOMETRY

In order to maintain reducing conditions for anaerobic microbial reactions that may be employed for bioremediation of chlorinated solvents, an electron donor must be available along with the appropriate alternate electron acceptor(s). Furthermore, the presence of a suitable electron donor is often necessary to prevent accumulation of chlorinated intermediates. These chemicals along with other major growth nutrients are often not among the chemical constituents available in the contaminated region and growth is limited without them. In order to stimulate anaerobic microbial growth and engineer a microbial treatment system for organic contaminant control, the chemical needs of the microorganisms must be defined and are given by the reaction stoichiometry.

Studies with methanogenic bacteria that biotransform chlorinated aliphatic compounds such as PCE, CF, and 1,1,1-TCA, using acetate as primary electron donor and carbon source, indicated the ratio of acetate mass used to the mass of chlorinated compounds transformed varied between 100/1 and 1000/1 (Bouwer and McCarty, 1983a). Recently, de Bruin et al. (1992) determined that 1 mM lactate was used for the complete dechlorination of 9 µM PCE to ethene. This amount of lactate is 150 times the minimum reducing equivalents necessary for a complete reduction of PCE to ethane. Consequently, large quantities of an electron donor like acetate or lactate may need to be injected into the contaminated soil system for anaerobic treatment of even a relatively small amount of chlorinated solvent contamination. However, de Bruin et al. (1992) did not attempt to optimize the lactate/PCE stoichiometry. The amount of lactate needed is likely to be markedly less (W. de Bruin, personal communication, 1992).

The appropriate amounts of electron donor, electron acceptor, and nutrients that must be supplied for growth of the anaerobic bacteria and the amounts of biomass and other products that will be formed can be estimated using the thermodynamic model reported by McCarty (1971). In this model, electrons from the electron donor can be coupled with the electron acceptor to generate energy or can be used to synthesize biomass. The relative amounts of the electron donor being oxidized for energy and being converted to biomass is established with an energy balance. The amount of energy released during oxidation of the electron donor must balance the amount of energy required to synthesize the cell material.

The balanced equations for the methanogenic system capable of complete reductive dechlorination of PCE to ethene as described by de Bruin et al. (1992) are given in Table 8.6. Such stoichiometric relationships established with the model can be used to determine the appropriate solution of lactate (electron donor) and nutrients to flush throughout the zone of contamination. The application of this stoichiometry for PCE soil contamination appears in Figure 8.2. When organic contaminants enter the subsurface, the residual amount retained on the soil after drainage has been found to range between 0.5% to 4% by volume depending on the soil type (Wilson, J. et al., 1986). Medium to fine sand typically retains 1 to 2% by volume of the organic contaminant. For this sand with bulk density of 2000 kg/m^3, about 16.3 kg of residual PCE (10 liters) will remain per m^3 of soil after free product recovery by pumping, and normal drainage leaves a residual of 1% by volume (Figure 8.2). According to the first set of reactions in Table 8.6, 970 kg of lactate and 17.4 kg of ammonia nitrogen would be required per m^3 of soil in order for the reductive dechlorination of PCE to ethene to proceed properly and be in balance. Other nutrients that are required in lesser quantities for bacterial growth are not included in the balanced equations. However, the phosphorus requirement is about one-sixth of that for nitrogen. Thus, this PCE biotransformation would require 2.9 kg of phosphorus per m^3 of soil. Bacterial biomass is represented by the empirical formula $C_5H_7O_2N$ and 140 kg of cells would be formed per m³ of soil during these anaerobic reactions. In this example, most of the reducing equivalents supplied as lactate evolve as methane (221 kg or 317 m³ of gas). The amounts of the innocuous products ethene and HCl formed from PCE are shown in Figure 8.2.

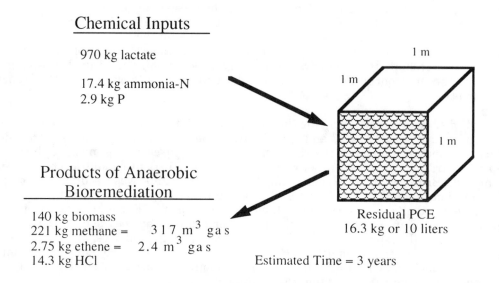

Chemical Inputs

970 kg lactate

17.4 kg ammonia-N
2.9 kg P

1 m

1 m

1 m

Products of Anaerobic Bioremediation

140 kg biomass
221 kg methane = 317 m³ gas
2.75 kg ethene = 2.4 m³ gas
14.3 kg HCl

Residual PCE
16.3 kg or 10 liters

Estimated Time = 3 years

Figure 8.2. **Chemical requirements and products of anaerobic bioremediation for one cubic meter soil contaminated with PCE using microbial system reported by de Bruin et al. (1992).**

As a result of the lactate addition for PCE biotransformation in the above example, 140 kg of anaerobic biomass would be formed. The microbial growth forms a larger mass than that of the original PCE retained by the soil. This biomass growth is likely to reduce the soil permeability and interfere with water flow during injection of chemicals. About 80% of the biomass will be biodegradable, and if additional oxygen or nitrate is supplied, eventually the biomass will decay to about 20% of the amounts given in Figure 8.2.

The mass balance relationships given above for lactate in a potential anaerobic bioremediation illustrate the enormous quantities of chemicals required and point to the urgent need to develop microbial systems with tighter coupling between the reducing equivalents from the electron donor and the reductive dechlorination reactions of the chlorinated solvent. A great improvement in the stoichiometry was recently reported by DiStefano et al. (1991) in a methanogenic system using methanol as primary electron donor. Nearly one-third of the methanol consumption was used for dechlorination of PCE to ethene. The resulting stoichiometry appears in Table 8.6. The corresponding chemical inputs and products per m³ of soil for the hypothetical PCE contamination introduced above appear in Figure 8.3. The amounts of methanol (13.6 kg) and nutrients (0.077 kg N and 0.013 kg P) necessary appear reasonable for field-scale applications. The amount of biomass formed (0.62 kg) per m³ of soil is not likely to cause problems with clogging near the injection well or permeability reduction in the formation.

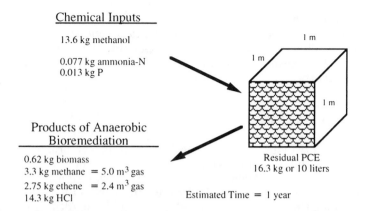

Chemical Inputs

13.6 kg methanol

0.077 kg ammonia-N
0.013 kg P

Products of Anaerobic
Bioremediation

0.62 kg biomass
3.3 kg methane $= 5.0 \text{ m}^3$ gas
2.75 kg ethene $= 2.4 \text{ m}^3$ gas
14.3 kg HCl

1 m

1 m

1 m

Residual PCE
16.3 kg or 10 liters

Estimated Time $= 1$ year

Figure 8.3. **Chemical requirements and products of anaerobic bioremediation for one cubic meter soil contaminated with PCE using microbial system reported by DiStefano et al. (1991).**

TABLE 8.6. STOICHIOMETRIC RELATIONSHIPS FOR POSSIBLE BIOREMEDIATION REACTIONS INVOLVING COMPLETE DECHLORINATION OF PCE TO ETHENE[a]

Stoichiometry for the Microbial System Reported by de Bruin et al. (1992)

PCE (C_2Cl_4) + 0.67 lactate $(C_3H_5O_3^-)$ + 2.33 H_2O ⟶
    ethene (C_2H_4) + 4 HCl + 1.33 CO_2 + 0.67 HCO_3^-

110 lactate + 12.6 NH_4^+ + 60 H_2O ⟶
    12.6 biomass $(C_5H_7O_2N)$ + 134 CH_4 + 97.9 HCO_3^- + 36.3 CO_2

Stoichiometry for the Microbial System Reported by DiStefano et al. (1991)[b]

PCE (C_2Cl_4) + 1.33 methanol (CH_3OH) + 1.33 H_2O ⟶
    ethene (C_2H_4) + 4 HCl + 1.33 CO_2

3.0 methanol + 2.25 HCO_3^- ⟶ 2.25 acetate (CH_3COO^-) + 0.75 CO_2 + 3.75 H_2O

2.25 acetate + 0.056 NH_4^+ + 2.02 H_2O + 0.083 CO_2 ⟶
    0.056 biomass + 2.11 CH_4 + 2.19 HCO_3^-

[a] All compounds were considered in aqueous phase except CO_2, CH_4, and ethene were taken as gaseous. Free energy of formation values for the organic compounds were obtained from Handbook of Organic Chemistry (Dean, 1987).

[b] The acetate produced from acetogenesis of methanol was assumed to undergo methane fermentation.

Another favorable biotransformation with an alternate electron acceptor is the conversion of CT to carbon dioxide under denitrification conditions as described in Table 8.1 (Criddle et al., 1990a). However, the relationship between acetate and nitrate consumption and CT biotransformation was not determined, thus further work is needed to clarify the stoichiometry.

8.9.   BIOTRANSFORMATION RATES

Knowledge of biotransformation rates is useful in a bioremediation process for determining the length of time required for meeting a treatment objective. Biotransformation half-lives observed for reductive dechlorination of chlorinated solvents in both environmental samples and the field range from weeks to months. For example, CF and other trihalomethanes were removed from the Palo Alto, California, Baylands Aquifer during injection of reclaimed municipal wastewater with half-lives of 3 to 6 weeks (Roberts et al., 1982). A much slower decline occurred in the concentrations of chlorinated ethanes and ethenes, yielding half-lives of 5 to 9 months. Similar anaerobic biotransformation rates for chlorinated solvents have been observed in sediment, aquifer microcosms, and in laboratory-scale batch and continuous-flow systems. Consequently, reductive dechlorination rates by indigenous microorganisms appear to be quite slow.

One of the objectives in a bioremediation scheme is to increase the numbers of desired microorganisms. Bouwer and Wright (1988) illustrate that elevation of the biomass concentration by one to two orders of magnitude by the addition of growth substrates and nutrients can correspondingly decrease the half-lives to between days and weeks. The lactate addition in the methanogenic system of de Bruin et al. (1992) yielded a high dechlorination rate of PCE to ethene (89 μmol/l/day). At this volumetric dechlorination rate, the anaerobic bioremediation of PCE illustrated in Figure 8.2 would require about 1,100 days or 3 years to complete. An even higher initial volumetric rate of PCE dechlorination (275 μmol/l/day) was obtained at 35°C with methanol as electron donor in the anaerobic system of DiStefano et al., (1991). At this volumetric dechlorination rate, the anaerobic bioremediation of PCE would be shortened to about 350 days or 1 year (illustrated in Figure 8.3).

In most studies of anaerobic biotransformation of chlorinated solvents, only 0.005 to 1.6% of the electrons available from the primary substrate (electron donor) were used for dechlorination. The transformations are believed to occur by cometabolism, and these cometabolic reactions are slow and inefficient due to the uncoupling of contaminant transformation and microbial growth. In cometabolism, enzymes produced by the microorganism to metabolize the primary substrate can interact with an organic contaminant and bring about its transformation in a fortuitous manner. The cometabolite does not provide energy for growth or maintenance, so the microorganism does not benefit from cometabolic transformations. Limited evidence exists that a halogenated compound can serve as sole energy and carbon source for anaerobic bacteria. Recently a homoacetogen has been isolated which used methyl chloride for growth (Traunecker et al., 1991). Another example is the novel bacterium PER-K23 already described (p. 8-7) which carries out respiration with H_2 as electron acceptor and PCE as electron donor (Holliger, 1992). The PCE reductive dechlorination rate for PER-K23 is several orders of magnitude higher than reaction rates observed for cometabolic reductive dechlorinations (Table 8.7). The coupling between reductive dechlorination and respiration is encouraging for developing anaerobic microbial systems with faster dechlorination rates that could be exploited for the bioremediation of sites contaminated with chlorinated solvents.

TABLE 8.7. PCE DECHLORINATION RATES BY DIFFERENT ANAEROBIC BACTERIA

Organism	Reaction	Product	Dechlorination Rate ($\mu mol/day/mg$ protein)	Reference[a]
PER-K23	respiration	*cis*-1,2-DCE	475	a
Acetobacterium woodii	cometabolism	TCE	0.086	b
Methanosarcina sp.	cometabolism	TCE	0.00084	c
Methanosarcina mazei	cometabolism	TCE	0.00048	c
Desulfomonile tiedjei	cometabolism	TCE	0.0023	c

[a]a = Holliger (1992); b = Egli et al. (1988); c = Fathepure et al. (1987)

Limited information is available on rates of CT biotransformation with nitrate as electron acceptor. CT was nearly completely biotransformed in batch microcosms after 3 weeks of incubation under denitrification conditions (Bouwer and McCarty, 1983b). CT was assimilated into cell mass, mineralized to CO_2, and reductively dechlorinated to CF, indicating simultaneous oxidative and reductive reactions. CT disappeared in a 10-day period with concomitant CF production in a denitrifying enrichment sample from Moffett Field, California (Criddle et al., 1990a). A field test of anaerobic bioremediation demonstrated that CT was transformed at rapid rates with half-lives on the order of hours to days through the addition of acetate in the presence of nitrate and sulfate (Semprini et al., 1991). CF was observed as an intermediate of the CT biotransformation. Although the rates of CT biotransformation reported in these studies are reasonably fast, the formation of CF as an intermediate product is objectionable. The denitrifying *Pseudomonas* sp. described on p. 8-7 was capable of complete biotransformation of CT without production of CF in 2 days. Further work is needed to evaluate if this reaction can be deployed in anaerobic bioremediation.

8.10. LIMITATIONS

The formation of chlorinated intermediates, biotransformation stoichiometry and rates, influence of mass transfer, and water quality changes under anaerobic conditions are major concerns and possible impediments to practical implementation of anaerobic bioremediation technology.

Incomplete reductive dechlorination of chlorinated solvents is often encountered under anaerobic conditions, which results in the formation and accumulation of lesser chlorinated aliphatic compounds. This reaction mechanism is probably the major obstacle to widespread deployment of anaerobic processes for in-situ bioremediation. The chlorinated compounds formed are objectionable and pose a threat to public health. An electron donor compound, such as lactate, methanol, or H_2, must be supplied to stimulate growth of the anaerobic microorganisms involved in the reductive dechlorination reactions. For this cometabolism, often the mass of primary substrate (electron donor) to mass of chlorinated solvent biotransformed ranges between 100/1 and 1000/1. Consequently, a large quantity of electron donor is needed along with additional nutrients like nitrogen and phosphorus. Proper delivery of such large masses of chemicals is an engineering challenge. The high levels of chemicals needed are converted to large amounts of end products, such as methane gas, carbon dioxide, and biomass. How to control these by-products, particularly in heterogeneous environments, is an important question yet to be resolved. The biomass growth is likely to fill up the pore space, causing marked permeability reduction in the formation. This plugging of the formation will in turn interfere with proper delivery of the chemicals required. Slow reaction rate is a final concern for the anaerobic microbial systems involved in reductive dechlorination. Half-lives for anaerobic reductive dechlorination are typically on the order of months in the field. Extrapolation of optimal rates presently observed in the laboratory suggests cleanup times of years in the field. The slow rates are likely to be problematic with most regulatory timetables.

The influence of mass transfer on the availability of chlorinated solvents for in-situ bioremediation is a potential drawback. Although the chlorinated solvents tend to weakly sorb to aquifer solids, as indicated by their small K_{ow} values, slow desorption can be the rate limiting step and control the bioavailability of the compounds. The practical effect of slow diffusion from within soil aggregates and other kinetic limitations to desorption is to decrease the rate of removal of the chlorinated solvents from the aquifer, thereby increasing the time required to achieve cleanup and the amount of chemicals that must be added to sustain anaerobic microbial activity. The ability to deliver electron donors, electron acceptors, and nutrients to the microorganisms is a second mass transfer problem. The effects of geologic complexity, such as strata of gravel, sand, silt and clay, and fractured layers, along with the difficulty of locating the sources of subsurface contamination, can severely hamper the supply of chemicals throughout the zone of contamination.

The intended use of the aquifer after treatment could create a problem with stimulating anaerobic processes. If the aquifer is to be a source of drinking water, then a number of negative water quality issues arise due to making the aquifer anaerobic. As conditions switch from oxic to anoxic, some metals will be solubilized, particularly iron and manganese. These metals cause taste and odors and stain materials that come in contact with the water (e.g. pipes, bathtubs, toilet bowls, sinks, and clothes). Metabolites excreted by the anaerobic biomass increase the organic matter content of the water. Disinfectants added to the water to control pathogens react with this organic matter to form disinfection by-products. Regulations under the Safe Drinking Water Act of 1986 are currently aimed at limiting the formation of such disinfection by-products. The addition of nitrate to aquifers to stimulate denitrifying conditions may be of concern because nitrate is regulated under the National Drinking Water Standards with a maximum contaminant level of 10.0 mg/l measured as nitrogen. Production of microbial metabolites can also solubilize cadmium, copper, lead, and zinc oxides (Francis and Dodge, 1988) which may facilitate their passage into the distribution system.

8.11. RESEARCH NEEDS

The positive attributes about bioremediation of chlorinated solvents under anaerobic conditions are that the compounds can be rendered harmless if reductive dechlorination is complete and costs are likely to be lower than other technologies. Additional research is needed to overcome the many limitations detailed above. Research needs specific for biotransformation of chlorinated solvents with alternate electron acceptors appear below:

1. There is need to determine the environmental factors and metabolic requirements necessary for either the complete reductive dechlorination of chlorinated solvent compounds or mineralization of certain compounds like CT to carbon dioxide. Achieving this goal requires further investigation into the physiology and metabolism of the various anaerobic microorganisms, focusing on both enriched mixed cultures and pure cultures. The aim of this research is to be able to specify completely and unambiguously the optimum conditions for growth and cometabolic transformation so that the biotransformation can be reliably controlled in the field.

2. It is essential to develop ways to minimize chemical requirements, particularly the need for an electron donor, and ways to increase reaction rates for the cometabolism of chlorinated solvents under anaerobic conditions. The kinetics of individual transformation processes must be thoroughly understood to enable confident prediction of intermediate and product formation and to permit design and scale-up of processes in the field.

3. The development of stable anaerobic microbial consortia that use reductive dechlorination for respiration rather than cometabolism would be beneficial for improving the success of anaerobic processes for in-situ bioremediation.

4   There is need to evaluate the feasibility of sequentially creating anaerobic followed by aerobic conditions in the subsurface in order to biotransform chlorinated solvents in a two step process. The impacts of the changing redox conditions on ground-water quality need to be assessed.

5   It is necessary to continue research to determine factors that govern the influence of sorption/desorption of chlorinated solvents on the performance of their bioremediation. If desorption is limiting, then ways must be devised to increase their availability for microbial transformation.

6. Optimal findings from the laboratory efforts mentioned above need to be demonstrated in the field by conducting small-scale studies at well-instrumented sites. Specific tasks to be conducted include evaluation of methods for delivery of chemicals (electron donors, electron acceptors, and nutrients) and their efficacy for stimulating anaerobic microorganisms, characterization of the extent of the biological active zone, and evaluation of contaminant residuals and times required for the microbial reactions.

8.12. CONCLUDING REMARKS

Chlorinated solvents are difficult to control in the environment; their widespread usage, uncontrolled disposal, and chemical/physical properties make them common ground-water contaminants.   The transformations of chlorinated solvents in the presence of alternate electron acceptors are important reactions that can affect their fate and can be applied to the development of treatment technology.   This section addressed some of the important factors that affect these biotransformations in the subsurface environment and that can be applied to the development of bioremediation technology. Many different anaerobic bacterial species are capable of catalyzing the reductive dechlorination of chlorinated solvents.   The conversion of CT to carbon dioxide under denitrification is another important anaerobic transformation.   An electron donor (primary substrate) is required to supply energy for bacterial growth and for activation of enzymes necessary for the transformation.   The possible conversion of chlorinated solvents to harmless products under anaerobic conditions has lead to an interest in using in-situ techniques for biotransformation of these contaminants as an alternative to aboveground treatment systems that generally involve physical/chemical processes that do not destroy the contaminants.   The objective of bioremediation is to stimulate the growth of indigenous or introduced microorganisms in regions of subsurface contamination, and thus, provide direct contact between microorganisms and the dissolved and sorbed contaminants for biotransformation.   The anaerobic process will typically require the perfusion of an electron donor, electron acceptor, and nutrients through the contaminated soil.   However, the supply of these chemicals can be difficult in tight and heterogeneous soils.   The formation of chlorinated intermediates, the large amounts of electron donor necessary for the cometabolic reductive dechlorination reactions, slow rates of desorption, and negative water quality changes are additional major concerns and possible impediments to practical implementation of anaerobic bioremediation technology.   Both basic laboratory studies and well-controlled field experiments are needed to establish feasibility and overcome the present limitations with exploiting anaerobic processes for bioremediation of chlorinated solvents.

ACKNOWLEDGEMENT

The author thanks Dr. Gosse Schraa, Department of Microbiology, Agricultural University, Wageningen, The Netherlands, for fruitful discussions on this topic.

REFERENCES

Bagley, D.M., and J.M. Gossett. 1990. Tetrachloroethene transformation to trichloroethene and *cis*-1,2-dichloroethene by sulfate-reducing enrichment cultures. *Appl. Environ. Microbiol.* 56(8):2511-2516.

Barrio-Lage, G., F.Z. Parsons, R.S. Nassar, and P.A. Lorenzo. 1986. Sequential dehalogenation of chlorinated ethenes. *Environ. Sci. Technol.* 20(1):96-99.

Belay, N., and L. Daniels. 1987. Production of ethane, ethylene, and acetylene from halogenated hydrocarbons by methanogenic bacteria. *Appl. Environ. Microbiol.* 53(7):1604-1610.

Bouwer, E.J., and P.L. McCarty. 1983a. Transformations of 1- and 2-carbon halogenated aliphatic organic compounds under methanogenic conditions. *Appl. Environ. Microbiol.* 45(4):1286-1294.

Bouwer, E.J., and P.L. McCarty. 1983b. Transformations of halogenated organic compounds under denitrification conditions. *Appl. Environ. Microbiol.* 45(4):1295-1299.

Bouwer, E.J., and J.P. Wright. 1988. Transformations of trace halogenated aliphatics in anoxic biofilm columns. *J. Contam. Hydrol.* 2:155-169.

Castro, C.E., and N.O. Belser. 1968. Biodehalogenation. Reductive dehalogenation of the biocides ethylene dibromide, 1,2-dibromo-3-chloropropane, and 2,3-dibromobutane in soil. *Environ. Sci. Technol.* 2(10):779-783.

Chemical and Engineering News. 1989. Facts and figures for the chemical industry at a glance. 67:(25):36-50.

Criddle, C.S., P.L. McCarty, M.C. Elliot, and J.F. Barker. 1986. Reduction of hexachloroethane to tetrachloroethylene in groundwater. *J. Contam. Hydrol.* 1:133-142.

Criddle, C.S., J.T. DeWitt, D. Grbic-Galic, and P.L. McCarty. 1990a. Reductive dechlorination of carbon tetrachloride by *Escherichia coli* K-12. *Appl. Environ. Microbiol.* 56(11):3247-3254.

Criddle, C.S., J.T. DeWitt, and P.L. McCarty. 1990b. Transformation of carbon tetrachloride by *Pseudomonas* sp. strain KC under denitrification conditions. *Appl. Environ. Microbiol.* 56:(11):3240-3246.

De Bruin, W.P., M.J.J. Kotterman, M.A. Posthumus, G. Schraa, and A.J.B. Zehnder. 1992. Complete biological reductive transformation of tetrachloroethene to ethane. *Appl. Environ. Microbiol.* 58(6):1966-2000.

Dean, J.A. 1987. *Handbook of Organic Chemistry*. New York: McGraw-Hill Book Company.

DiStefano, T.D., J.M. Gossett, and S.H. Zinder. 1991. Reductive dechlorination of high concentrations of tetrachloroethene to ethene by an anaerobic enrichment culture in the absence of methanogenesis. *Appl. Environ. Microbiol.* 57(8):2287-2292.

Dooley-Danna, M., S. Fogel, and M. Findlay. 1989. The sequential anaerobic/aerobic biodegradation of chlorinated ethenes in an aquifer simulator. In: *Proceedings International Symposium on Processes Governing the Movement and Fate of Contaminants in the Subsurface Environment.* Stanford University. Stanford, California. July 23-26.

Egli, C., R. Scholtz, A.M. Cook, and T. Leisinger. 1987. Anaerobic dechlorination of tetrachloromethane and 1,2-dichloroethane to degradable products by pure cultures of *Desulfobacterium* sp. and *Methanobacterium* sp. *FEMS Microbiol. Lett.* 43:257-261.

Egli, C., T. Tschan, R. Scholtz, A.M. Cook, and T. Leisinger. 1988. Transformation of tetrachloromethane to dichloromethane and carbon dioxide by *Acetobacterium woodii.* *Appl. Environ. Microbiol.* 54(11):2819-2824.

Egli, C., S. Stromeyer, A.M. Cook, and T. Leisinger. 1990. Transformation of tetra- and trichloromethane to CO_2 by anaerobic bacteria is a non-enzymic process. *FEMS Microbiol. Lett.* 68:207-212.

Fathepure, B.Z., J.P. Nengu, and S.A. Boyd. 1987. Anaerobic bacteria that dechlorinate perchloroethene. *Appl. Environ. Microbiol.* 53(11):2671-2674.

Fathepure, B.Z., and S.A. Boyd. 1988. Dependence of tetrachloroethylene dechlorination on methanogenic substrate consumption by *Methanosarcina* sp. strain DCM. *Appl. Environ. Microbiol.* 54(12):2976-2980.

Francis, A.J., and C.J. Dodge. 1988. Anaerobic microbial dissolution of transition and heavy metal oxides. *Appl. Environ. Microbiol.* 54(4):1009-1014.

Freedman, D.L., and J.M. Gossett. 1989. Biological reductive dechlorination of tetrachloroethylene and trichloroethylene to ethylene under methanogenic conditions. *Appl. Environ. Microbiol.* 55(9):2144-2151.

Freeze, R.A., and J.A. Cherry. 1979. *Ground Water.* Englewood Cliffs, New Jersey. Prentice-Hall, Inc.

Galli, R., and P.L. McCarty. 1989. Biotransformation of 1,1,1-trichloroethane, trichloromethane, and tetrachloromethane by a *Clostridium* sp. *Appl. Environ. Microbiol.* 55(4):837-844.

Holliger, C. 1992. *Reductive Dehalogenation by Anaerobic Bacteria.* Ph.D. Thesis. Department of Microbiology, Agricultural University. Wageningen, The Netherlands.

Kästner, M. 1991. Reductive dechlorination of tri- and tetrachloroethylenes depends on transition from aerobic to anaerobic conditions. *Appl. Environ. Microbiol.* 57(7):2039-2046.

Klecka, G.M., S.J. Gonisor, and D.A. Markham. 1990. Biological transformation of 1,1,1-trichloroethane in subsurface soils and ground water. *Environ. Toxicol. Chem.* 9:1437-1451.

Kloepfer, R.D., D.M. Easley, B.B. Haas, Jr., T.G. Deihl, D.E. Jackson, and C.J. Wurrey. 1985. Anaerobic degradation of trichloroethylene in soil. *Environ. Sci. Technol.* 19(3):277-280.

Krone, U.E., K. Laufer, R.K. Thauer, and H.P.C. Hogenkamp. 1989. Coenzyme F_{430} as a possible catalyst for the reductive dehalogenation of chlorinated C_1 hydrocarbon in methanogenic bacteria. *Biochemistry.* 28:10061-10065.

Lyman, W.J., W.F. Reehl, and D.H. Rosenblatt. 1982. *Handbook of Chemical Property Estimation Methods.* McGraw-Hill. New York, NewYork.

Mackay, D.M., P.V. Roberts, and J.A. Cherry. 1985. Transport of organic contaminants in groundwater. *Environ. Sci. Technol.* 19(5):384-392.

Mackay, D.M., and J.A. Cherry. 1989. Groundwater contamination: Pump-and-treat remediation. *Environ. Sci. Technol.* 23(6):630-636.

Major, D.W., E.W. Hodgins, and B.J. Butler. 1991. Field and laboratory evidence of in situ biotransformation of tetrachloroethene to ethene and ethane at a chemical transfer facility in North Toronto. In: *On-Site Bioreclamation Processes for Xenobiotic and Hydrocarbon Treatment.* Eds., R.E. Hinchee and R.F. Olfenbuttel. Butterworth-Heinemann. Boston, Massachusetts. pp. 147-171.

McCarty, P.L. 1971. Energetics and bacterial growth. In: *Organic Compounds in Aquatic Environments.* Eds., J. Faust and J.V. Hunter. Marcel Dekker, Inc. New York. pp. 495-531.

Mikesell, M.D., and S.A. Boyd. 1990. Dechlorination of chloroform by *Methanosarcina* strains. *Appl. Environ. Microbiol.* 56(4):1198-1201.

Parsons, F., P.R. Wood, and J. DeMarco. 1984. Transformation of tetrachloroethene and trichloroethene in microcosms and groundwater. *J. Amer. Water Works Assoc.* 72(2):56-59.

Parsons, F., and G.B. Lage. 1985. Chlorinated organics in simulated groundwater environments. *J. Amer. Water Works Assoc.* 77(5):52-59.

Parsons, F., G.B. Lage, and R. Rice. 1985. Biotransformation of chlorinated organic solvents in static microcosms. *Environ. Toxicol. Chem.* 4:739-742.

Pye, V.I., R. Patrick, and J. Quarles. 1983. *Groundwater Contamination in the United States.* University of Pennsylvania Press. Philadelphia, Pennsylvania.

Roberts, P.V., J. Schreiner, and G.D. Hopkins. 1982. Field study of organic water quality changes during ground water recharge in the Palo Alto Baylands. *Water Res.* 16(6):1025-1035.

Roberts, P.V., G.D. Hopkins, D.M. Mackay, and L. Semprini. 1990. A field evaluation of in-situ biodegradation of chlorinated ethenes: Part 1, Methodology and field site characterization. *Ground Water.* 28(4):591-604.

Scholtz-Muramatsu, H., R. Szewzyk, U. Szewzyk, and S. Gaiser. 1990. Tetrachloroethylene as electron acceptor for the anaerobic degradation of benzoate. *FEMS Microbiol. Lett.* 66:81-86.

Semprini, L., G.D. Hopkins, P.V. Roberts, and P.L. McCarty. 1991. In situ biotransformation of carbon tetrachloride, freon-113, freon-11, and 1,1,1-TCA under anoxic conditions. In: *On-Site Bioreclamation Processes for Xenobiotic and Hydrocarbon Treatment*. Eds., R.E. Hinchee and R.F. Olfenbuttel. Butterworth-Heinemann. Boston, Massachusetts. pp. 41-58.

Sewell, G.W., and S.A. Gibson. 1991. Stimulation of the reductive dechlorination of tetrachloroethene in anaerobic aquifer microcosms by the addition of toluene. *Environ. Sci. Technol.* 25(5):982-984.

Stumm, W., and J.J. Morgan. 1981. *Aquatic Chemistry*. 2nd Ed. John Wiley and Sons, Inc. New York, New York.

Thauer, R.K., K. Jundermann, and K. Decker. 1977. Energy conservation in chemotrophic anaerobic bacteria. *Bacteriol. Rev.* 41:100-180.

Thomas, J.M., and C.H. Ward. 1989. In situ biorestoration of organic contaminants in the subsurface. *Environ. Sci. Technol.* 23(7):760-766.

Traunecker, J., A. Preuss, and G. Diekert. 1991. Isolation and characterization of a methyl chloride utilizing strictly anaerobic bacterium. *Arch. Microbiol.* 156:416-421.

U.S. Environmental Protection Agency. 1985. *Chemical, Physical, and Biological Properties of Compounds Present at Hazardous Waste Sites*. EPA 530/SW-89/010. Office of Solid Waste and Emergency Response Directive 9850.3. NTIS PB89-132203. September, 1985. 543 pp.

Verschueren, K. 1983. *Handbook of Environmental Data on Organic Chemicals, 2nd Edition*. Van Nostrand Reinhold Co., Inc. New York, New York.

Vogel, T.M., and P.L. McCarty. 1985. Biotransformation of tetrachloroethylene to trichloroethylene, dichloroethylene, vinyl chloride, and carbon dioxide under methanogenic conditions. *Appl. Environ. Microbiol.* 49(5):1080-1083.

Vogel, T.M., and P.L. McCarty. 1987. Abiotic and biotic transformations of 1,1,1-trichloroethane under methanogenic conditions. *Environ. Sci. Technol.* 21(12):1208-1213.

Vogel, T.M., B.Z. Fathepure, and H. Selig. 1989. Sequential anaerobic/aerobic degradation of chlorinated organic compounds. In: *International Symposium on Processes Governing the Movement and Fate of Contaminants in the Subsurface Environment*. Stanford University. Stanford, California. July 23-26.

Wagner, K., K. Boyer, R. Claff, M. Evans, S. Henry, V. Hodge, S. Mahmud, D. Sarno, E. Scopino, and P. Spooner. 1986. *Remedial Action Technology for Waste Disposal Sites, 2nd Edition*. Noyes Data Corporation. Park Ridge, New Jersey.

Wilson, B.H., G.B. Smith, and J.F. Rees. 1986. Biotransformations of selected alkylbenzenes and halogenated aliphatic hydrocarbons in methanogenic aquifer material: A microcosm study. *Environ. Sci. Technol.* 20(10):997-1002.

Wilson, J.T., L.E. Leach, M. Henson, and J.N. Jones. 1986. In situ biorestoration as a groundwater remediation technique. *Ground Water Monit. Rev.* 6(4):56-64.

Yang, J., and R.E. Speece. 1986. The effects of chloroform toxicity on methane fermentation. *Water Res.* 20(10):1273-1279.

SECTION 9

NATURAL BIOREMEDIATION OF HYDROCARBON-CONTAMINATED GROUND WATER

Robert C. Borden
Civil Engineering Department
North Carolina State University
Raleigh, North Carolina 27695-7908
Telephone: (919)515-2331
Fax: (919)515-7908

9.1. GENERAL CONCEPT OF NATURAL BIOREMEDIATION

The basic concept behind "Natural Bioremediation" is to allow naturally occurring microorganisms to degrade contaminants that have been released into the subsurface and at the same time minimize risks to public health and the environment. Use of this approach will require an assessment of those factors that influence the biodegradation capacity of an aquifer and the potential human and environmental risks. Ongoing research has shown that an aquifer's assimilative capacity depends on the metabolic capabilities of the native microorganisms, the aquifer hydrogeology and geochemistry, and the contaminants involved.

Natural bioremediation is not a "No Action" alternative. In most cases, natural bioremediation is used to supplement other conventional remediation techniques. The type and extent of conventional remediation techniques used depend on the environmental conditions in the aquifer, the extent of contamination, and the risk to the public and environment. In some cases, only removal of the primary source (e.g., leaking tanks, contaminated soil) may be necessary. In other situations, conventional ground-water remediation by pump and treat may be used to reduce the concentrations within the aquifer. Once contaminant concentrations are reduced below some defined level, the pump and treat system may be terminated and natural bioremediation used to complete the cleanup.

Implementation of a natural remediation system differs from conventional techniques, in that a portion of the aquifer is allowed to remain contaminated. Depending on site specific conditions, use of natural bioremediation may require a variance from existing regulations, may involve questions of third party liability and property rights, or require public hearings and review by elected officials. Natural bioremediation is less predictable than conventional pump and treat or excavation. Consequently, some type of risk evaluation will usually be required whenever natural bioremediation is considered.

The purpose of this chapter is to discuss the potential for natural bioremediation to be incorporated into an overall remedial design at a hazardous waste site. The various biological processes using oxygen, nitrate, ferric iron, sulfate, and carbonate as electron acceptors will be addressed. Considered also are the effect of environmental conditions on biodegradation, site characterization needed for natural bioremediation, necessary parameters to be monitored, performance and prediction of natural bioremediation, and issues that may affect the costs associated with the technology. Well-documented case studies of natural bioremediation at former wood preserving facilities and petroleum releases are also presented.

At present, there is almost no operating history to judge the effectiveness of natural bioremediation. Early attempts at aquifer remediation focused on using conventional remediation techniques to remove or permanently immobilize contaminants at the highest priority sites. While at many low priority sites, regulators may have assumed that natural bioremediation would be adequate to control the migration of dissolved contaminants, typically these sites have not been monitored sufficiently to determine if this approach is actually effective or to identify those factors that influence the efficiency of natural bioremediation. At present, there are no well-documented full scale demonstrations of natural bioremediation, although there has been some limited research into the processes that control the natural biodegradation of dissolved hydrocarbon plumes (Borden et al., 1986; Barker et al., 1987; Franks, 1987; Hult, 1987a; Chiang et al., 1989; Wilson et al, 1993). At present, the primary repositories of expertise on natural bioremediation are in universities, in industry, the U.S. Environmental Protection Agency (U.S. EPA) and the U.S. Geological Survey (U.S.G.S.). Use of natural bioremediation is typically much less expensive than other remediation technologies. Consequently, there has been less incentive for private consultants and service companies to invest funds developing this technique.

9.2. HYDROCARBON DISTRIBUTION, TRANSPORT AND BIODEGRADATION IN THE SUBSURFACE

Dissolved hydrocarbons are among the most common ground-water contaminants and can originate from spilled fuels (gasoline, diesel, jet fuel, heating oil), solvents, wood preservatives, and coal gasification wastes. Many of these wastes are initially present as nonaqueous phase liquids (NAPLs), which contain many different components. Gasoline contains primarily the lighter, lower boiling point compounds like pentane and benzene; while creosote and coal tars contain more of the higher boiling point compounds.

Ground water that comes in contact with the residual hydrocarbons will dissolve a portion of the NAPL. The amount of each individual component that dissolves in water can be roughly estimated as the aqueous solubility times the mole fraction of the individual component in the oily phase. Ground waters that have come in contact with petroleum fuels typically become contaminated with BTEX compounds (benzene, toluene, ethylbenzene, and xylenes) because these compounds are the most water soluble. Ground water in contact with gasoline will typically contain more BTEX than ground water in contact with heating oil and other heavier hydrocarbons, because gasoline contains a higher percentage of BTEX. These ground waters may also contain high concentrations of fuel additives since many of these additives are highly soluble in water and are present in relatively high concentrations in some gasolines.

Once an individual hydrocarbon constituent is dissolved, it may be transported by moving ground water. While the movement of most petroleum constituents will be retarded to some extent by sorption onto aquifer material, the more soluble compounds are usually not sorbed to a large extent, except in aquifers with a high organic carbon content. The primary mechanisms that will limit the subsurface transport of dissolved hydrocarbons are biodegradation and, to a lesser extent, volatilization. Volatilization results in a transfer of the lower boiling point, more volatile compounds from the ground water to the soil gas within the unsaturated zone. At present, the significance of this process is unknown, although it is expected that the relative importance of volatilization will be much less for large spills where a larger portion of the plume is present at significant depth below the water table. Nonbiological (abiotic) reactions such as hydrolysis are of lesser importance because many hydrocarbons are relatively stable under the environmental conditions found in most aquifers.

9.2.1. Petroleum Hydrocarbon Biodegradation

Hydrocarbon biodegradation can be represented by the chemical reaction

Hydrocarbon + electron acceptors + microorganisms + nutrients →
→ carbon dioxide + water + microorganisms + waste products

The rate and extent of hydrocarbon biodegradation in the subsurface will depend on several factors, including (1) the quantity and quality of nutrients and electron acceptors; (2) the type, number and metabolic capability of the microorganisms; and (3) composition and amount of the hydrocarbons. While virtually all petroleum hydrocarbons are biodegradable, the rate and extent of biodegradation can be highly variable. Depending on environmental conditions, biodegradation may be very rapid or very slow.

9.2.2. Subsurface Microorganisms

Recent studies have shown that an active diverse microbial population exists in the subsurface, often at great depth. The organisms present appear to be predominantly bacteria, but some fungi and protozoa have been identified (Ghiorse and Wilson, 1988). The native organisms appear to be well adapted to low nutrient conditions. Many of the organisms identified grow very poorly or not at all under high nutrient conditions, yet thrive at low levels of organic carbon (Ghiorse and Balkwill, 1985). Biochemical analyses have indicated the presence of storage granules, allowing survival during extended starvation periods (White et al., 1983).

Most of the organisms identified are aerobes, but strict anaerobes have been identified from a few sites (Ghiorse and Wilson, 1988). Microbially mediated denitrification was observed in a sand and gravel aquifer contaminated with treated sewage (Smith and Duff, 1988). Anaerobic bacteria were identified by van Beelen and Fleuren-Kemila (1989) from two sandy aquifers, a saturated peat soil and a river sediment. Chapelle et al. (1987) identified methanogenic and sulfate-reducing bacteria from sediments collected 20 to 180 m below grade in the Maryland coastal plain. Recent work by Jones et al. (1989) has shown that methanogens are present in the subsurface at over 300 m below grade in sediments near Aiken, SC. Although the microbial community was dominated by aerobic microorganisms, sulfate-reducing and methanogenic organisms could be identified from most sediments throughout the depth profile. In most cases, the total number of methanogens were very low, but the organisms present were capable of degrading a wide variety of organic substrates, e.g., benzoate, phenol, lactate, formate, acetate.

The ability of microorganisms to degrade a wide variety of hydrocarbons is well known. In an early review, Zobell (1946) identified over 100 microbial species from 30 genera that could degrade some type of hydrocarbon. In a more recent study, Ridgeway et al. (1990) identified 309 gasoline-degrading bacteria from a shallow coastal aquifer contaminated with unleaded gasoline. Hydrocarbon-degrading microorganisms are widespread in the environment and occur in fresh and salt water, soil, and ground water. Litchfield and Clark (1973) analyzed ground-water samples from 12 different aquifers throughout the United States that were contaminated with hydrocarbons. These workers found hydrocarbon- utilizing bacteria in all samples at densities up to 1.0×10^6 cells per ml. After a gasoline spill in Southern California, McKee et al. (1972) found 50,000 hydrocarbon-degrading bacteria per ml or higher in samples from wells containing traces of gasoline, while a noncontaminated well had only 200 organisms per ml.

9.2.3. Use of Different Electron Acceptors For Biodegradation

Hydrocarbon biodegradation is essentially an oxidation-reduction reaction where the hydrocarbon is oxidized (donates electrons) and an electron acceptor (e.g., oxygen, nitrate, etc.) is reduced (accepts electrons). There are a number of different compounds that can act as electron acceptors including oxygen (O_2), nitrate (NO_3^-), iron oxide (e.g. $Fe(OH)_3$), sulfate ($SO_4^=$), and carbon dioxide (CO_2). Aerobic bacteria can only use molecular oxygen (O_2) as an electron acceptor. Anaerobic bacteria use other compounds such as NO_3^-, $SO_4^=$, $Fe(OH)_3$, or CO_2 as electron acceptors. Oxygen is the most preferred electron acceptor because microorganisms gain more energy from aerobic reactions. Sulfate and carbon dioxide are the least preferred because microorganisms gain the least energy from these reactions.

9.2.3.1. Aerobic Biodegradation

Almost all petroleum hydrocarbons are biodegradable under aerobic conditions. Oxygen is a cosubstrate for the only known enzyme that can initiate the metabolism of hydrocarbon (Young, 1984) and is later used as an electron acceptor for energy generation. Under ideal conditions, biodegradation rates for low molecular weight aliphatic, olefinic and aromatic compounds can be very high. Alvarez and Vogel (1991) observed essentially complete removal of mixtures of benzene, toluene, and *p*-xylene in aquifer slurries and pure cultures after 3 to 13 days incubation. Using aquifer material from a gas plant facility in Michigan, Chiang et al. (1989) found between 80 and 100% removal of BTX (120-16,000 ppb) in microcosms with sufficient oxygen. Half-lives for biodegradation varied between 5 and 20 days. In a field test of aerobic biodegradation at a former wood preserving facility, the total polynuclear aromatic hydrocarbon concentration dropped by over 90% within 24 hours of the start of the test when sufficient oxygen was available (Borden et al., 1989).

The ease of biodegradation will depend somewhat on the type of hydrocarbon. Moderate to lower molecular weight hydrocarbons (C_{10} to C_{24} alkanes, single ring aromatics) appear to be the most easily degradable hydrocarbons (Atlas, 1988). As the molecular weight increases, so does the resistance to biodegradation. Gasoline contains primarily the low to moderate molecular weight compounds while diesel and coal tars contain more of the higher molecular weight compounds. Jamison et al. (1975) found that the vast majority of gasoline components were readily degraded by a mixed microbial population obtained from a gasoline contaminated aquifer. Many of the individual gasoline components would not support microbial growth as a sole carbon source but did disappear when gasoline dissolved in water was used as the substrate. This suggested that a mixed microbial population may be necessary for complete degradation. In a study of the catabolic activity of bacteria from an aquifer contaminated with unleaded gasoline, Ridgeway et al. (1990) found that most isolates were very specific in their ability to degrade hydrocarbons. Although all of the 15 hydrocarbons tested were degraded by at least one isolate, most organisms were able to degrade only one of several closely related compounds. Toluene, *p*-xylene, ethylbenzene, and 1,2,4-trimethylbenzene were most frequently utilized, whereas cyclic and branched alkanes were least utilized.

In many cases, the major limitation on aerobic biodegradation in the subsurface is the low solubility of oxygen in water. For example, aerobic toluene biodegradation can be represented by the theoretical reaction:

$$C_6H_5\text{-}CH_3 + 9\,O_2 \xrightarrow{\text{bacteria}} 7\,CO_2 + 4\,H_2O + \text{Energy} \tag{1}$$

Water saturated with air contains from 6 to 12 mg/l of dissolved oxygen. Complete conversion of toluene (and many other hydrocarbons) to carbon dioxide and water requires approximately 3 grams of oxygen per gram of hydrocarbon. Using this ratio, the oxygen present in water could result in the biodegradation of 2 to 4 mg/l of dissolved hydrocarbon by strictly aerobic processes. If the hydrocarbon concentration is greater than this, biodegradation may be incomplete or may occur via slower anaerobic processes.

9.2.3.2. Biodegradation via Nitrate Reduction

When the oxygen supply is depleted and nitrate is present (or other oxidized forms of nitrogen), some facultatively anaerobic microorganisms will utilize nitrate (NO_3^-) as a terminal electron acceptor instead of oxygen. For toluene, this process can be approximated by the theoretical reaction:

$$C_6H_5\text{-}CH_3 + 6\,NO_3^- \xrightarrow{\text{bacteria}} 7\,CO_2 + 4\,H_2O + 3\,N_2 + \text{Energy} \tag{2}$$

Over the past decade, researchers have found that toluene, ethylbenzene, *m*-, *p*-, and *o*-xylene, naphthalene and a variety of other compounds can be biodegraded using nitrate as the terminal electron acceptor (Kuhn et al., 1985; Zeyer et al., 1986; Kuhn et al., 1988; Hutchins et al., 1991; Mihelcic and Luthy, 1991). At this time, there is some question about the biodegradability of benzene under denitrifying conditions. Several investigators have reported benzene to be recalcitrant (not biodegradable) under denitrifying conditions (Kuhn et al., 1988; Zeyer et al.; 1990; Hutchins et al., 1991) whereas other studies indicate that benzene is degraded (Major et al., 1988; Kukor and Olsen, 1989).

9.2.3.3. Biodegradation Using Ferric Iron

Once the available oxygen and nitrate are depleted, subsurface microorganisms may use oxidized ferric iron [Fe(III)] as an electron acceptor. Microorganisms have been identified that can couple the reduction of ferric iron with the oxidation of aromatic compounds including toluene, phenol, *p*-cresol and benzoate (Lovley and Lonergan, 1990; Lovley et al., 1989). Large amounts of ferric iron are present in the sediments of most aquifers and could potentially provide a large reservoir of electron acceptor for hydrocarbon biodegradation. This iron may be present in both crystalline and amorphous forms. The forms that are most easily reduced are amorphous and poorly crystalline Fe(III) hydroxides, Fe(III) oxyhydroxides, and Fe(III) oxides (Lovley, 1991). A possible reaction coupling the oxidation of toluene to the reduction of Fe(III) in ferric hydroxide [$Fe(OH)_3$] can be approximated as:

$$C_6H_5\text{-}CH_3 + 36\,Fe(OH)_3 \xrightarrow{\text{bacteria}}$$
$$7\,CO_2 + 36\,Fe^{+2} + 72\,OH^- + 22\,H_2O + \text{Energy} \tag{3}$$

The reduction of Fe(III) may result in high concentrations (10 to 100 mg/l) of dissolved Fe(II) in contaminated aquifers. Lovley et al. (1989) found that in an aquifer contaminated by a crude oil spill, the selective removal of benzene, toluene and xylenes from the plume was accompanied by an accumulation of dissolved Fe(II) and depletion of Fe(III) oxides in the contaminated sediments. Although the exact mechanism of microbial ferric iron reduction is poorly understood, the available evidence suggests that iron reduction is an important mechanism in the subsurface biodegradation of dissolved hydrocarbons.

9.2.3.4. Biodegradation via Sulfate Reduction and Methanogenesis

Past research has shown that a wide variety of problem organics may be biodegraded by sulfate-reducing and/or methanogenic (methane generating) microorganisms (Grbic-Galic, 1990). These compounds include creosol isomers (Smolenski and Suflita, 1987), homocyclic and heterocyclic aromatics (Berry et al., 1987), alkylbenzenes (Grbic-Galic and Vogel, 1987; Beller et al. 1991), and unsaturated hydrocarbons (Schink, 1985). Sulfate reducers could potentially biodegrade toluene using sulfate in the following theoretical reaction (Beller et al., 1991):

$$C_6H_5\text{-}CH_3 + 4.5 \ SO_4^= + 3 \ H_2O \xrightarrow{\text{bacteria}}$$
$$2.25 \ H_2S + 2.25 \ HS^- + 7 \ HCO_3^- + 0.25 \ H^+ + Energy \qquad (4)$$

Methanogenic consortia (groups of microorganisms which generate methane) could potentially biodegrade toluene using water as an electron acceptor (Vogel and Grbic-Galic, 1986) in the following theoretical reaction:

$$C_6H_5\text{-}CH_3 + 5 \ H_2O \xrightarrow{\text{bacteria}} 4.5 \ CH_4 + 2.5 \ CO_2 + Energy \qquad (5)$$

At this time, little is known about the effect of sulfate reduction and methanogenic biodegradation on the fate of dissolved hydrocarbons in the subsurface. While there are well-documented reports of toluene biodegradation via sulfate reduction (Beller et al., 1991) and methanogenesis (Grbic-Galic and Vogel, 1987), the extent and significance of hydrocarbon biodegradation using these electron acceptors is poorly understood. This may be partially due to the characteristics of these microorganisms. Sulfate-reducing and methanogenic consortia are known to be very sensitive to a variety of environmental conditions including temperature, inorganic nutrients (nitrogen, phosphorus, trace metals), toxicants, and pH (Zehnder, 1978). An imbalance in any of these factors could significantly reduce the rate and extent of biodegradation.

9.2.4. Effect of Environmental Conditions on Biodegradation

In most cases, the environmental factor that has the greatest influence on the rate and extent of biodegradation is the availability of suitable electron acceptors (oxygen, nitrate, etc.). In addition to electron acceptor concentration, temperature, pH, and nutrients can influence biodegradation.

The optimum temperature for growth of most microorganisms present in shallow aquifers is between 25°C and 40°C. In northern portions of the continental U.S., shallow ground-water temperatures can be as low as 3°C and could significantly reduce the growth rate of subsurface microorganisms. This lower growth rate may be somewhat offset by the higher solubility of oxygen in water at lower temperatures. In the central and southern U.S., ground-water temperatures are higher (15°C to 25°C) and should not significantly impair biodegradation.

The optimum pH for microbial growth is dependent on the specific microorganisms and their respiration pathways. Aerobic microorganisms often tolerate a wider range in pH, whereas many anaerobes are sensitive to pH and operate efficiently only in a narrow pH range. Denitrification and methanogenic biodegradation rates are usually optimum between pH 7 and 8, and may drop off rapidly below pH of 6 (van den Berg, 1974; U.S. EPA, 1975). The pH of most water supply aquifers is between 6.0 and 8.5, although waters having lower pH are not uncommon (Hem, 1989).

The primary nutrients required for microbial growth are nitrogen, phosphorus, sulfur and low levels of various minerals (Fe, Mn, etc.). Dissolution of the parent rock typically releases some minerals and hence it is unlikely that these nutrients would be completely absent (McNabb and Dunlap, 1975). Depending on the extent of microbial growth, one or more of these nutrients may become limiting. In enhanced bioremediation projects, nitrogen and phosphorus are frequently added to allow maximum growth (Lee et al., 1988). In passive remediation systems, the extent of microbial natural growth will be much lower and nutrient limitations will probably be less severe. Lee and Ward (1984) found that addition of nitrogen, phosphorus and trace minerals increased bacterial growth in creosote contaminated ground water but did not increase the extent of contaminant removal.

9.3. NATURAL BIOREMEDIATION OF A HYDROCARBON PLUME

While there are no truly typical sites, it may be useful to consider a hypothetical site where the aquifer hydrogeology and geochemistry are reasonably well defined. For this hypothetical case, assume that a small release of gasoline has occurred from an underground storage tank (UST). The soils immediately below the tank are contaminated with moderate levels of residual hydrocarbon. A simple schematic of this site is shown in Figure 9.1.

Rainfall infiltrating through the hydrocarbon-contaminated soils will leach out some of the more soluble hydrocarbon components, probably benzene, toluene, ethylbenzene and xylenes (BTEX); fuel additives, methyl tertiary butyl ether (MTBE) and ethylene dibromide (EDB); and a smaller portion of the less soluble constituents (aliphatic hydrocarbons and higher molecular weight aromatics). As the hydrocarbon-contaminated water migrates downward through the unsaturated zone, a portion of the dissolved hydrocarbons may biodegrade. The extent of biodegradation will be controlled by the size of the spill and the rate of downward movement. For larger spills, the available oxygen will be consumed and aerobic biodegradation will not continue.

Dissolved hydrocarbons that do not completely biodegrade within the unsaturated zone will be carried downward, enter the saturated zone and be transported downgradient within the water table aquifer. Figure 9.2 shows a simple schematic plan view of a dissolved hydrocarbon plume undergoing biodegradation. Near the source area, dissolved hydrocarbons enter the saturated zone and flow downgradient. Native microorganisms will use the available oxygen in the source area to biodegrade a portion of the hydrocarbon. Dissolved hydrocarbons that are not biodegraded will then be carried downgradient in a plume of anaerobic contaminated water. In this region, the extent of biodegradation will probably be limited by the available oxygen supply. Because the solubility of oxygen in water is relatively low, only a small amount of hydrocarbon may be biodegraded aerobically in the source area.

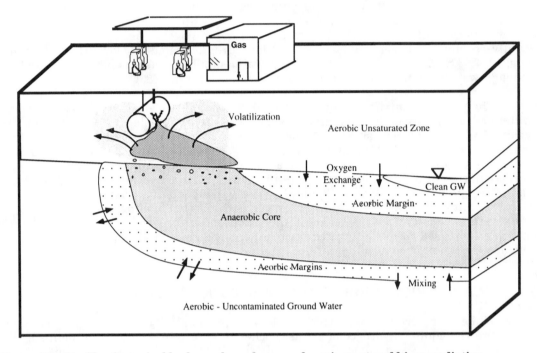

Figure 9.1.   Profile of a typical hydrocarbon plume undergoing natural bioremediation.

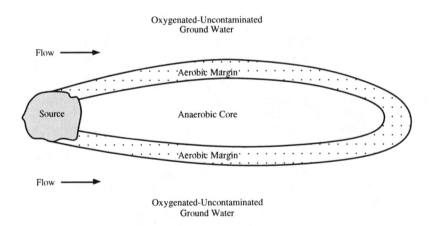

Figure 9.2.   Plan view of a typical hydrocarbon plume undergoing natural bioremediation.

As the plume migrates downgradient, dispersion will mix the anaerobic hydrocarbon-contaminated water with clean oxygenated water at the plume fringes. This is the region where most aerobic biodegradation occurs. After an acclimation period, a population of aerobic hydrocarbon-degrading bacteria will develop in the sediments of this fringe area. As oxygenated water mixes with hydrocarbon contaminated water, the attached bacteria will consume both the hydrocarbon and oxygen, preventing the further spread of the contaminant plume. This is the reason why many dissolved hydrocarbon plumes appear long and narrow. As the dissolved hydrocarbons disperse outward, they come in contact with oxygenated ground water and biodegrade. If this process is allowed to continue indefinitely, the dissolved hydrocarbon plume will reach a quasi-steady state condition where the long-term rate of hydrocarbon dissolution is equal to the rate of biodegradation. The major limitations to this process are the amount of oxygen present and the extent of mixing. Recent field studies (Freyberg, 1986; Moltyaner and Killey, 1988) have shown that mixing (dispersion) in most aquifers is very limited and consequently, the overall rate of aerobic biodegradation may be slow.

Hydrocarbon biodegradation using nitrate will probably follow the same general pattern as aerobic biodegradation. Nitrate present in the uncontaminated ground water will mix with hydrocarbon at the plume fringes and increase the amount of biodegradation. While oxygen is preferred to nitrate by most microorganisms, nitrate is still a good electron acceptor and many hydrocarbons can be biodegraded in this manner (e.g. toluene, ethylbenzene, xylenes). Also, any hydrocarbons biodegraded using nitrate will not express a demand for oxygen, which will allow further aerobic biodegradation of other hydrocarbons that require oxygen.

In the core of the plume, conditions may become highly reducing and other anaerobic biodegradation reactions may occur. Certain hydrocarbons and many bacterial waste products may be biodegraded by iron reduction, sulfate reduction or methanogenic biodegradation. While little is known about these processes, it is clear that they do occur. Field monitoring has shown that in the core of some hydrocarbon plumes, sulfate concentrations are reduced and dissolved iron and methane concentrations are elevated (Wilson et al., 1990). It is not yet known whether these conditions result from direct anaerobic attack by bacteria on the hydrocarbon molecule or anaerobic biodegradation of bacterial waste products. From a practical perspective, it may not matter. Any organic carbon (hydrocarbons or waste products) biodegraded by an anaerobic pathway will reduce the total oxygen demand on the aquifer. By reducing the overall oxygen demand, more oxygen will be available for those compounds that can only be biodegraded aerobically.

9.4. CASE STUDIES OF NATURAL BIOREMEDIATION

Over the past ten years, there have been a number of well-documented studies which have demonstrated that plumes of dissolved hydrocarbons will biodegrade in the subsurface without human intervention. These studies have included former wood preserving facilities and petroleum releases.

One of the earliest studies of natural bioremediation was conducted at the United Creosoting Company site in Conroe, Texas, by a team of researchers from the R.S. Kerr Environmental Research Laboratory (U.S. EPA) and the National Center for Ground Water Research. Early work (Lee and Ward, 1984; Wilson et al., 1985) demonstrated that an adapted population of creosote-degrading microorganisms was present within the contaminated zone but not in the uncontaminated regions of the aquifer. Later studies correlated creosote biodegradation with the availability of dissolved oxygen (Lee and Ward, 1984). These results were used to develop and calibrate the computer model, BIOPLUME, to simulate hydrocarbon transport and aerobic biodegradation within the aquifer (Borden and Bedient, 1986; Borden et al., 1986). Model results indicated that removal of the contaminant source would be sufficient to contain the hydrocarbon plume and active remediation by pump and treat would not be required.

Microbiologists from the U.S. Geological Survey have studied two different creosote-contaminated aquifers where methanogenic degradation of organic compounds has been observed. Field studies at a contaminated aquifer in St. Louis Park, Minnesota, showed that methane production was occurring in zones within the aquifer that had been contaminated with creosote (Godsy et al., 1983). Later studies demonstrated that the presence of anaerobes (denitrifiers, iron reducers, sulfate reducers and methanogens) was highly correlated with the presence of creosote. More recent work at an abandoned creosote plant in Pensacola, Florida, has shown a wide variety of organic compounds present in the aquifer were undergoing methanogenic biodegradation and that transport distances in the aquifer could be correlated with biodegradation rates observed in laboratory microcosms (Troutman et al., 1984; Goerlitz et al., 1985).

Monitoring at petroleum contamination sites suggests that methanogenic biotransformation of petroleum related compounds may be more common than has generally been assumed. Ehrlich et al. (1985) observed elevated numbers of sulfate-reducing and methanogenic bacteria in a jet fuel contaminated aquifer. Evans and Thompson (1986) and Marrin (1987) monitored methane concentrations in soil gas to map subsurface hydrocarbon contamination. In a study of soil gas concentrations near underground storage tanks, Payne and Durgin (1988) found elevated methane concentrations at over 20% of the 36 sites surveyed. Methane gas production can be so rapid that safety problems occur at some sites. Hayman et al. (1988) had to develop a special apparatus to remove the large quantities of methane generated from a fuel spill at the Miami, Florida, airport.

Hult (1987b) observed the production of large volumes of methane in the unsaturated zone immediately below a crude oil spill at the U.S. Geological Survey research site in Bemidji, Minnesota. At this same site, Eganhouse et al. (1987) observed a two order of magnitude decrease in alkylbenzene concentration over a downgradient travel distance of 150 m. This decrease was accompanied by elevated concentrations of aliphatic and aromatic acids in the ground water (Baedecker et al., 1987). The acids included benzoic, methylbenzoic, trimethylbenzoic, toluic, cyclohexanoic, and dimethylcyclohexanoic. These are the same acids identified by Grbic-Galic and Vogel (1987) as intermediates in anaerobic degradation of alkylbenzenes. Ground-water and sediment analyses demonstrated that methanogenic biodegradation caused a drop in pH and a rise in bicarbonate concentrations in the ground water. The actual drop in ground-water pH appears to have been limited by dissolution of carbonate minerals (and possibly aluminosilicates) (Siegel, 1987).

9.5.   SITE CHARACTERIZATION FOR NATURAL BIOREMEDIATION

The first step in evaluating a site for potential application of natural bioremediation is to complete a conventional site characterization. This characterization should include: (1) detailed description of the subsurface hydrology and geology; (2) delineation of the contaminant source area and any mobile NAPLs; (3) delineation of the horizontal and vertical extent of the contaminant plume; and (4) identification of any downgradient receptors (wells or surface discharges) that could potentially be affected. In some cases it may be appropriate to model ground-water flow and/or transport at the site to gain a better understanding of the hydrologic system and contaminant transport pathways.

In addition to the basic data required in most remedial investigations, information will be needed to evaluate the ability of the aquifer to assimilate wastes and the potential risks if the system does not perform as expected. Specific questions that should be addressed are described below.

9.5.1. Is the Contaminant Biodegradable?

The first major question to be addressed is whether the contaminants are biodegradable by microorganisms present at the site. The level of detail required to answer this question will depend on the type of contaminant and general site conditions.

The most common dissolved hydrocarbons (benzene, toluene, ethylbenzene and xylenes) released from gasoline spills are known to be readily biodegradable under aerobic conditions (Jamison et al., 1975; Gibson and Subramanian, 1984; Thomas et al., 1990; Alvarez and Vogel, 1991). In addition, aerobic hydrocarbon-degrading microorganisms are very common in nature and have been recovered from virtually all petroleum-contaminated sites that have been studied (Litchfield and Clark, 1973). For most petroleum sites, extensive studies to confirm the presence of BTEX-degrading microorganisms are probably not necessary. In contrast to BTEX, there is much less information available on the biodegradability of many fuel additives such as methyl tertiary butyl ether (MTBE), 1,2-dibromoethane (EDB), or 1,2-dichloroethane (EDC). If persistence of fuel additives is a concern, site specific studies may be needed to confirm the presence of microorganisms capable of degrading these compounds and to estimate biodegradation rates.

Ground water contaminated with creosote, coal tar and heavier petroleum products often contains higher molecular weight aromatic compounds (fluorene, phenanthrene, dibenzofuran, etc.). These compounds often biodegrade much more slowly and may persist for long time periods even under ideal conditions (Lee, 1986; Borden et al., 1989). Site specific laboratory studies may be needed to determine if these compounds are biodegradable by subsurface microorganisms and if the rates of biodegradation are sufficient to contain the contaminant plume.

9.5.2. Is Biodegradation Occurring in the Aquifer?

Probably the most important question to address is whether the compounds of concern are actually biodegrading in the aquifer. The simplest way to answer this question is to examine the ground-water monitoring data and determine if there is a significant decline in the total mass of the contaminant as the plume migrates downgradient. Unfortunately, it is often difficult to evaluate changes in total mass without an extensive monitoring well network. Comparison of dissolved hydrocarbon concentrations at individual points is not sufficient to prove biodegradation, since dispersion will reduce the point concentrations even if there is no biodegradation. To overcome these problems, other parameters are often used as secondary indicators of biodegradation.

One very useful method for assessing the extent of biodegradation is to monitor changes in the concentration of inorganic compounds within the aquifer. Biodegradation of dissolved hydrocarbons will result in the removal of electron acceptors (oxygen, nitrate, and sometimes sulfate) and release of waste products (carbon dioxide and sometimes reduced iron and methane) in areas where microorganisms are most active. If field monitoring indicates that oxygen, nitrate and/or sulfate is being depleted (or carbon dioxide, soluble iron, or methane is being produced) within the plume, this is a good indication that one or more of the contaminants are being biodegraded. The major limitation of this approach is that it is not possible to determine which specific compounds are being degraded.

A second method that can be used to determine if individual compounds are being biodegraded is to examine changes in the ratio of different contaminants along the flow path. If one contaminant declines more rapidly than another, this suggests that some process is removing that contaminant. Field monitoring at several hydrocarbon plumes (Jasiorkowski and Robbins, 1991) and a sanitary landfill leachate plume (Barker et al., 1986) has shown a more rapid downgradient decline in *o*-xylene concentrations than *m*- or *p*-xylene. Since all of the xylene isomers should sorb to the aquifer equally, the only explanation for this pattern would be biodegradation of the *o*-xylene.

9.5.3. Are Environmental Conditions Appropriate for Biodegradation?

As previously discussed, virtually all hydrocarbons are biodegradable. Yet extensive plumes of dissolved hydrocarbons persist in some aquifers. Why does this apparent contradiction occur? The answer lies with the environmental conditions in the specific aquifer.

Virtually all hydrocarbons biodegrade more rapidly in the presence of dissolved oxygen. If dissolved oxygen concentrations are low in a specific aquifer, the rate of natural biodegradation will be lower. Also, the pH of the aquifer should be near neutrality, adequate inorganic nutrients should be present (nitrogen, phosphorus, and trace minerals), and no toxicants should be present that could inhibit microbial growth.

In most cases, it is not necessary to perform extensive investigations to precisely determine the concentrations of nitrogen, phosphorus, trace minerals, and potential toxicants. Past studies have shown that most aquifers do not contain toxicants and do contain adequate levels of inorganic nutrients to support moderate levels of microbial growth (Lee, 1986). If field monitoring indicates that biodegradation is occurring, it can reasonably be assumed that aquifer conditions are appropriate for microbial growth. Where field monitoring data suggest that biodegradation is being inhibited, additional laboratory studies may be needed to identify those factors that are limiting biodegradation. When performing laboratory studies, it is very important to design the experiment to simulate actual conditions within the aquifer. For example, if the oxygen supply in the aquifer is limiting, laboratory studies conducted with an excess of oxygen (or nitrogen, phosphorus, etc.) will overestimate the actual extent of biodegradation and lead to erroneous conclusions.

9.5.4. If the Waste Doesn't Completely Biodegrade, Where Will It Go?

Natural bioremediation, like other available techniques, is not foolproof. Instances arise where for some unforeseen reason, the contaminant plume does not biodegrade as expected. In order to adequately manage a natural remediation system, it is first necessary to evaluate the consequences of a system failure. In most cases, the primary consequences of a failure will be: (1) contamination of water supply wells; or (2) contamination of surface water. Appropriate controls should be incorporated into a natural remediation system to identify a failure and eliminate it.

9.6. MONITORING NATURAL BIOREMEDIATION SYSTEMS

One of the most important factors to consider in planning a natural bioremediation system is monitoring system performance. The monitoring system typically includes: (1) interior wells to monitor the actual plume distribution and indicator parameters; and (2) guardian wells at the outside edge of the area of contamination to monitor potential offsite migration and determine if additional remedial measures are required.

Interior wells may be monitored to evaluate the overall system performance. Parameters to be monitored typically include: (1) individual hydrocarbon components; (2) dissolved oxygen; (3) nitrate; (4) dissolved iron; (5) redox potential; (6) carbon dioxide; (7) pH; and (8) total organic carbon. Monitoring of individual hydrocarbon components can be performed using standard techniques and provides an indication of the treatment effectiveness.

Dissolved oxygen is monitored to determine if one or more of the organics are biodegrading and as an aid in defining the contaminant plume. Typically, both dissolved oxygen and hydrocarbon concentrations will be reduced at the margins of the plume. Dissolved oxygen can be measured in the field using electrodes or field test kits. Collection of accurate data on dissolved oxygen concentrations in ground water is difficult because of problems with aerating the samples during collection. One to two mg/l of oxygen may be added to the sample during collection unless special precautions are taken to prevent aeration. The extent of aeration can be reduced by using special pumps and filling the well casing with argon gas, but in most cases aeration cannot be completely eliminated.

Nitrate and iron may be monitored to determine the extent of anaerobic biodegradation of the hydrocarbons and any bacterial waste products. Nitrate can be monitored by collecting samples using conventional techniques and then transporting to the laboratory for analysis. Collection of samples for iron analysis is more difficult because of problems with iron present in suspended solids and precipitation of dissolved iron during transport. One method that may be used is to filter samples in the field during collection, preserve them with a concentrated acid and then analyze for total iron. While this procedure does not differentiate between dissolved ferric and ferrous iron, in most cases essentially all iron in excess of 0.5 mg/l will be in the reduced ferrous form (Hem, 1989).

Measurement of redox potential is relatively simple and can provide a good qualitative indicator of the overall oxidation-reduction status of the aquifer. Redox potential can be measured using a platinum electrode and a standard pH meter. In locations where the redox potential is negative, the ground water is strongly reduced, indicating significant bacterial decomposition. In areas where the redox potential is positive, the ground water is oxidizing, indicating that the contaminant plume has not reached this point or that bacterial degradation has not occurred. In most cases, redox potentials should not be used for precise calculations but as a qualitative indicator of environmental conditions within and outside the contaminant plume (Barcelona et al., 1989).

Carbon dioxide and pH can be monitored to evaluate the extent of bacterial respiration and determine if conditions are suitable for biodegradation. If the pH falls outside of a specified range (typically 5 to 9), biodegradation may be inhibited. Accumulation of carbon dioxide within and adjoining the contaminant plume is indicative of bacterial respiration. Direct interpretation of carbon dioxide concentrations is sometimes difficult because of shifts in the dominant form of inorganic carbon with pH and release of inorganic carbon during dissolution of certain minerals.

Individual hydrocarbon components can be monitored to determine the extent of the contaminant plume and any organic waste products produced during biodegradation of the dissolved hydrocarbons. In some cases, dissolved hydrocarbons will not be completely biodegraded but will be converted to nontoxic organic waste products. Monitoring total organic carbon (TOC) will provide some indication of the total oxygen demand exerted by the contaminant plume.

Guardian wells may be installed at the outside edge of the contamination area to monitor system performance, evaluate the potential for offsite migration and determine if additional remedial measures are required. In most cases, these wells are used for regulatory purposes and are only monitored for compounds of regulatory concern. These wells may also be monitored occasionally for indicator parameters (oxygen, nitrate, etc.) to confirm that the wells do in fact intercept the 'plume' of ground water that has undergone biodegradation.

9.7. PERFORMANCE OF NATURAL BIOREMEDIATION SYSTEMS

Under optimal conditions, natural bioremediation should be capable of completely containing a dissolved hydrocarbon plume. While there are few well-documented cases where this has occurred, there is a great deal of anecdotal evidence that suggests that natural bioremediation can be effective in containing dissolved hydrocarbon plumes. Typically greater than 90% of all underground tanks are used to store gasoline and other petroleum fuels. Yet a study by the California Department of Health Services (Hadley and Armstrong, 1991) found that by far the most common ground-water contaminants were chlorinated solvents, not petroleum constituents. These results suggest that the petroleum contaminants are being removed to below detection limits before reaching water supply wells.

In many aquifers, conditions will not be perfect for natural bioremediation and less than optimal biodegradation will occur. The extent of aerobic biodegradation will be controlled by the amount of contamination released, the rate of oxygen transfer into the subsurface, and the background oxygen content of the aquifer. When large amounts of contamination enter the subsurface, they overwhelm the capacity of an aquifer to assimilate them. As a result, extensive contamination may persist for long distances. When hydrogeologic conditions such as clayey, confining layers or naturally occurring organic deposits reduce the rate of oxygen transfer into the subsurface, the assimilative capacity of the aquifer will be lower. Anaerobic biodegradation may be inhibited by low pH, low buffering capacity, or absence of appropriate electron acceptors (nitrate, iron, etc.). Heterogeneous conditions within the aquifer may prevent mixing and allow a portion of the plume to migrate rapidly. If this occurs, the extent of biodegradation may be less than would be expected for more uniform conditions.

9.8. PREDICTING THE EXTENT OF NATURAL BIOREMEDIATION

One of the most frequently asked questions is "How far will the plume migrate before it biodegrades?" Unfortunately, this is a very difficult question to answer.

To predict the maximum extent of plume migration, it is necessary to estimate: (1) the rate of migration; and (2) the rate of biodegradation. The rate of contaminant migration can be estimated by measuring the hydraulic gradient and permeability of the aquifer. Accurate estimation of the biodegradation rate within an aquifer is much more difficult. Results from laboratory studies may significantly over- or underestimate biodegradation rates if environmental conditions in the laboratory differ from conditions in the field.

Computer models may be used to combine the results of field and laboratory investigations and to predict the actual extent of biodegradation in an aquifer. At present, there are a number of computer models that have been developed to simulate contaminant biodegradation (Molz et al., 1986; Odencrantz et al., 1989; MacQuarrie et al., 1990). Two of the most commonly used models for simulating hydrocarbon biodegradation are: (1) first-order decay models; or (2) BIOPLUME II.

Kemblowski et al. (1987) describe the use of a first order decay model to simulate hydrocarbon biodegradation at several sites. Their results show that this simple approach can adequately match the observed hydrocarbon distribution in the aquifers studied. The major limitation of this method is in estimating the first-order decay rate before extensive data are collected. Once the contaminant plume is properly delineated and shown to be biodegrading, it is possible to match the field data to a first-order decay equation and estimate the decay rate.

Hydrocarbon biodegradation may also be simulated using the computer model BIOPLUME II (Rifai et al., 1989). BIOPLUME II is based on the U.S.G.S. Method of Characteristics model (Konikow and Bredehoeft, 1978) and includes advection, dispersion, oxygen-limited biodegradation, and first-order decay in a two-dimensional aquifer. Oxygen-limited biodegradation is simulated as an instantaneous reaction between oxygen and hydrocarbon. Calibration of BIOPLUME II is relatively simple because the only data required are the aquifer hydrogeology, background oxygen concentrations and contaminant source concentrations.

The major limitations of BIOPLUME II are the inability to accurately simulate dissolution of residual hydrocarbons and anaerobic biodegradation of hydrocarbons or bacterial waste products. BIOPLUME II assumes that all contaminants are converted directly to carbon dioxide and water using 3 mg of oxygen for every mg of hydrocarbon degraded. In many cases, this significantly underestimates the amount of biodegradation (Chiang et al., 1989) and leads to a conservative prediction. This error is presumably due to anaerobic degradation of bacterial waste products and certain hydrocarbons. Anaerobic decay can be simulated in BIOPLUME II using a first-order decay rate, but this approach suffers from the same limitation as the simple first-order decay models. There are no accurate methods available to estimate these decay rates without first collecting extensive field data.

In summary, there are no good methods available at this time for predicting the extent of hydrocarbon biodegradation without first characterizing the contaminant plume. Once the contaminant plume is defined, there are several methods that can be used to analyze the available data and evaluate the effect of different alternatives on contaminant migration. As additional field data becomes available from different sites, it may become possible to estimate the decay rate by extrapolating results from similar aquifers and avoid extensive field data collection.

9.9.   ISSUES THAT MAY AFFECT THE COSTS OF THIS TECHNOLOGY

One of the major factors controlling the costs of natural bioremediation is acceptance of this approach by regulators, environmental groups and the public. At sites where natural bioremediation is strongly opposed, the costs of implementation may actually be higher than conventional remediation technologies (e.g. pump and treat). In North Carolina, regulations have been in place for five years that allow responsible parties to request a reclassification of contaminated ground water to a nonwater supply use. Once reclassified, the responsible party would not be required to actively remediate the site. At present there are over 2,000 sites under investigation, with over 100 pump-and-treat systems in operation. Yet no one has ever filed a request for reclassification. The apparent cause is a perception by the responsible parties that the legal, administrative and site characterization costs for reclassification would be excessive, and the probability of success would be low.

The second major issue limiting application of natural bioremediation is third party liability. A hydrocarbon plume that is left in place to naturally biodegrade may migrate under an adjoining property, posing a potential risk to public health and the environment. Even when public health is not at risk, adjoining property owners may have strong concerns about a contaminant plume migrating under their property and the potential impact on property values. In such cases, natural bioremediation could be coupled with active plume management technology, such as purge wells, to prevent undesirable impact to third parties.

9.10.   KNOWLEDGE GAPS AND RESEARCH OPPORTUNITIES

Currently, there are no reliable methods for predicting the effectiveness of natural bioremediation without first conducting extensive field work. Existing mathematical models cannot be used in a predictive mode because they either: (1) require extensive field data for calibration; or (2) greatly underestimate the extent of anaerobic biodegradation. This is often the primary reason why natural bioremediation is not seriously considered when evaluating remedial alternatives. Without some reasonable assurance of success, responsible parties are not willing to risk the large sums of money required for legal, administrative and site characterization costs.

Over the next several years, there is potential to dramatically improve our ability to predict the extent of natural bioremediation. Several organizations [U.S. EPA, American Petroleum Institute (API), Electric Power Research Institute (EPRI)] are funding extensive field studies to characterize dissolved hydrocarbon plumes undergoing natural bioremediation. These studies will generate an extensive database that will be used to improve our understanding of the basic processes that control natural biodegradation and to develop more accurate models for predicting the extent of natural bioremediation. In order to use this database effectively, additional research is needed in two general areas: (1) anaerobic hydrocarbon biodegradation; and (2) biodegradation modeling.

We now know that many hydrocarbons can be biodegraded under anaerobic conditions using nitrate, iron, sulfate, water and carbon dioxide as terminal electron acceptors. What we do not know is what factors control the rate of anaerobic hydrocarbon biodegradation and why anaerobic hydrocarbon biodegradation occurs in some locations and not in others. Detailed laboratory studies are needed to resolve these questions.

Primary emphasis should be placed on coordinating these laboratory studies with the ongoing field work to maximize benefits.

Existing models of hydrocarbon biodegradation do not adequately represent anaerobic biodegradation. Consequently, these models grossly under-predict the extent of biodegradation at many sites. Until this problem is resolved, natural bioremediation will not be seriously considered at many sites where it is a reasonable alternative. The extensive field database being collected by EPA, API and EPRI provides an outstanding opportunity to resolve this problem. By coordinating model development with the field data collection, in the next few years we can significantly improve our ability to predict the extent of natural bioremediation.

REFERENCES

Alvarez, P.J.J., and T.M. Vogel. 1991. Substrate interactions of benzene, toluene, and para-xylene during microbial degradation by pure cultures and mixed culture aquifer slurries. *Appl. Environ. Microbiol.* 57(10):2981-2985.

Atlas, R.M. 1988. *Microbiology: Fundamentals and Applications*. 2nd Edition. MacMillan Publish. Co. New York, New York. p. 457.

Barcelona, M.J., T.R. Holm, M.R. Schock, and G.K. George. 1989. Spatial and temporal gradients in aquifer oxidation-reduction conditions. *Water Resourc. Res.* 25(5):991-1003.

Baedecker, M.J., I.M. Cozzarelli, and J.A. Hopple. 1987. *The Composition and Fate of Hydrocarbons in a Shallow Glacial-Outwash Aquifer*. U.S. Geological Survey Open File Report 87-109. pp. C23-C24.

Barker, J.F., J.S. Tessman, P.E. Plotz, and M. Reinhard. 1986. The organic geochemistry of a sanitary landfill leachate plume. *J. Contaminant Hydrol.* 1:171-189.

Barker, J.F., G.C. Patrick, and D. Major. 1987. Natural attenuation of aromatic hydrocarbons in a shallow sand aquifer. *Ground Water Monitoring Review.* 7(1):64-71.

Beller, H.R., E.A. Edwards, D. Grbic-Galic, and M.Reinhard. 1991. *Microbial Degradation of Alkylbenzenes Under Sulfate-Reducing and Methanogenic Conditions*. EPA/600/2-91/027.

Berry, D.F., A.J. Francis, and J.M. Bollag. 1987. Microbial metabolism of homocyclic and heterocyclic aromatic compounds under anaerobic conditions. *Microbiol. Reviews.* 51(1):43-59.

Borden, R. C., and P. B. Bedient. 1986. Transport of dissolved hydrocarbons influenced by reaeration and oxygen limited biodegradation: 1. Theoretical development. *Water Resourc. Res.* 22(1):1973-1982.

Borden, R. C., P. B. Bedient, M. D. Lee, C. H. Ward, and J. T. Wilson. 1986. Transport of dissolved hydrocarbons influenced by reaeration and oxygen limited biodegradation: 2. Field application. *Water Resourc. Res.* 22(1):1983-1990.

Borden, R. C., M. D. Lee, J.M. Thomas, P. B. Bedient, and C. H. Ward. 1989. In situ measurement and numerical simulation of oxygen limited biodegradation. *Ground Water Monitoring Review.* 9(1):83-91.

Chapelle, F.H., J.L. Zelibor, D.J. Grimes, and L.L. Knobel. 1987. Bacteria in deep coastal plain sediments of Maryland: A possible source of CO_2 to groundwater. *Water Resour. Res.* 23(8):1625-1632.

Chiang, C.Y., J.P. Salanitro, E.Y. Chai, J.D. Colthart and C.L. Klein. 1989. Aerobic biodegradation of benzene, toluene, and xylene in a sandy aquifer - Data analysis and computer modeling. *Ground Water.* 27(6):823-834.

Eganhouse, R.P., T.F. Dorsey, C.S. Phinney. 1987. *Transport and Fate of Monoaromatic Hydrocarbons in the Subsurface at the Bemidji, Minnesota, Research Site*. U.S. Geological Survey Open File Report 87-109. pp. C29-C30.

Ehrlich, G.G., R.A. Schroeder, and P. Martin. 1985. *Microbial Populations in a Jet-Fuel-Contaminated Shallow Aquifer at Tustin, California*. U.S. Geological Survey Open File Report 85-335. 14 p.

Evans, O.D., and G.M. Thompson. 1986. Field and interpretation techniques for delineating subsurface petroleum hydrocarbon spills using soil gas. In: *Proceedings of Petroleum Hydrocarbons and Organic Chemicals in Ground Water: Prevention, Detection and Restoration*. National Water Well Association. Dublin, Ohio. pp. 444-455.

Franks, B.J. 1987. Introduction, Chapter A. *Movement and Fate of Creosote Waste in Ground Water Near an Abandoned Wood-Preserving Plant near Pensacola, Florida*. U.S. Geological Survey Open File Report 87-109. pp. A3-A10.

Freyberg, D.L. 1986. A natural gradient experiment on solute transport in a sand aquifer 2. Spatial moments and the advection and dispersion of nonreactive tracers. *Water Resourc. Res*. 22(13):2031-2046.

Ghiorse, W.C., and D.L. Balkwill. 1985. Microbial characterization of subsurface environments. In: *Ground Water Quality*. Eds., C.H. Ward, W. Gieger, and P.L. McCarty. John Wiley and Sons, Inc. New York, New York. pp. 387-401.

Ghiorse, W.C., and J.T. Wilson. 1988. Microbial ecology of the terrestrial subsurface. *Adv. Appl. Microbiol*. 33:107-172.

Gibson, D.T., and V. Subramanian. 1984. Microbial degradation of aromatic hydrocarbons. In: *Microbial Degradation of Organic Compounds*. Ed., D.T. Gibson. Marcel Dekker, Inc. pp. 181-252.

Godsy, E.M., D.F. Goerlitz, and G.G. Ehrlich. 1983. Methanogenesis of phenolic compounds by a bacterial consortium from a contaminated aquifer in St. Louis Park, Minnesota. *Bull. Environ. Contam. Toxicol*. 30:261-268.

Goerlitz, D.F., D.E. Troutman, E.M. Godsy, and B.J. Franks. 1985. Migration of wood preserving chemicals in contaminated groundwater in a sand aquifer at Pensacola, Florida. *Environ. Sci. Technol*. 19(10):955-961.

Grbic-Galic, D., and T. M. Vogel. 1987. Transformation of toluene and benzene by mixed methanogenic cultures. *Appl. Environ. Microbiol*. 53(2):254-260.

Grbic-Galic, D. 1990. Anaerobic microbial transformation of nonoxygenated aromatic and alicyclic compounds in soil, subsurface and freshwater sediments. In: *Soil Biochemistry*. Eds., J.M. Bollag and G. Stotzky. Marcel Dekker, Inc. New York. pp. 117-189.

Hadley, P.W., and R. Armstrong. 1991. "Where's the benzene?" - Examining California ground-water quality surveys. *Ground Water*. 29(1):35-40.

Hayman, J.W., R.B. Adams, and J.J. McNally. 1988. Anaerobic biodegradation of hydrocarbons in confined soils beneath busy places: A unique problem of methane control. In: *Proceedings of Petroleum Hydrocarbons and Organic Chemicals in Ground Water: Prevention, Detection and Restoration*. National Water Well Association. Dublin, Ohio. pp. 383-396.

Hem, John D. 1989. *Study and Interpretation of the Chemical Characteristics of Natural Waters*. U.S. Geological Survey Water Supply Paper 2254. 263 p.

Hult, M.F. 1987a. *Microbial Oxidation of Petroleum Vapors in the Unsaturated Zone*. U.S. Geological Survey Open File Report 87-109. pp. C25-C26.

Hult, M.F. 1987b. Introduction, Chapter C. *Movement and Fate of Crude Oil Contaminants on the Subsurface Environment at Bemidji, Minnesota*. U.S. Geological Survey Open File Report 87-109. pp. C3-C6.

Hutchins S.R., W.C. Downs, J.T. Wilson, G.B. Smith, and D.K. Kovacs. 1991. Effect of nitrate addition on biorestoration of fuel-contaminated aquifer field demonstration. *Ground Water*. 29(4):571.

Jamison, V.W., R.L. Raymond, and J.O. Hudson Jr. 1975. Biodegradation of high-octane gasoline in groundwater. *Dev. Ind. Microbiol*. 16:305-311.

Jasiorkowski, J.L., and G.A. Robbins. 1991. Systematic variations in relative abundances of aromatic compounds in gasoline contaminated ground water. In: *Proceedings of Focus Conference on Eastern Regional Ground Water Issues*. National Water Well Association. Dublin, Ohio. pp. 769-781.

Jones, R.E., R.E. Beeman, J.M. Suflita. 1989. Anaerobic metabolic processes in the deep terrestrial subsurface. *Geomicrobiol. J*. 7:117-130.

Kemblowski, M.W., J.P. Salanitro, G.M. Deeley, and C.C. Stanley. 1987. Fate and transport of residual hydrocarbon in groundwater - A case study. In: *Proceedings of Petroleum Hydrocarbons and Organic Chemicals in Ground Water: Prevention, Detection and Restoration*. National Water Well Association. Dublin, Ohio. pp. 207-231.

Konikow, L.F., and J.D. Bredehoeft. 1978. Computer model of two-dimensional solute transport and dispersion in ground water. *Automated Data Processing and Computations, Techniques of Water Resources Investigations of the U.S. Geological Survey*. Book 7, Chapter C2. Washington, DC. 90 p.

Kuhn E.P., P.J. Colberg, and J.L. Schnoor. 1985. Microbial transformation of substituted benzene during infiltration of river water to groundwater: Laboratory column studies. *Environ. Sci. Technol*. 19(10):961-968.

Kuhn E.P., J. Zeyer, P. Eicher, and R.P. Schwarzenbach. 1988. Anaerobic degradation of alkylated benzene in denitrifying laboratory aquifer columns. *Appl. Environ. Microbiol*. 54(2):490-496.

Kukor J.J., and R.H. Olsen. 1989. Diversity of toluene degradation following long-term exposure to BTEX in situ. *Biotechnology and Biodegradation*. Portfolio Publishing. The Woodlands, Texas. pp. 405-421.

Lee, M.D. 1986. *Biodegradation of Organic Contaminants at Hazardous Waste Disposal Sites*. Ph.D. Dissertation. Rice University. Houston, Texas. 160 p.

Lee, M.D., and C.H.Ward. 1984. Microbial ecology of a hazardous waste site: Enhancement of biodegradation. In: *Proceedings Second International Conference on Ground Water Quality Research*. Oklahoma State University. Stillwater, Oklahoma. pp. 25-27.

Lee, M.D., R.C. Borden, J.T. Wilson, M. Thomas, P.B. Bedient, and C.H.Ward. 1988. Biorestoration of organic contaminated aquifers. *CRC Critical Reviews in Environmental Control*. 18(1):629-636.

Litchfield, J.H., and L.C. Clark. 1973. *Bacterial Activities in Ground Waters Containing Petroleum Products*. American Petroleum Institute. Pub. No. 4211.

Lovley, D.R., M.J. Baedecker, D.J. Lonergan, I.M. Cozzarelli, E.J.P. Phillips, and D.I. Siegel. 1989. Oxidation of aromatic contaminants coupled to microbial iron reduction. *Nature*. 339(6222):297-299.

Lovley, D.R., and D.J. Lonergan. 1990. Anaerobic oxidation of toluene, phenol, and p-cresol by the dissimilatory iron-reducing organisms, GS-15. *Appl. Environ. Microbiol*. 56(6):1858-1864.

Lovley, D.R. 1991. Dissimilatory Fe(III) and Mn(IV) reduction. *Microbiol. Reviews*. 55(2):259-287.

MacQuarrie, K.T.B., E.A. Sudicky, and E.O. Frind. 1990. Simulation of biodegradable organic contaminants in groundwater 1. Numerical formulation and model calibration. *Water Resourc. Res*. 26(2):207-222.

Major D. W., C.I. Mayfield, and J.F. Baker. 1988. Biotransformation of benzene by denitrification in aquifer sand. *Ground Water*. 26(1):8-14.

Marrin, D.L. 1987. Soil gas analysis of methane and carbon dioxide: delineating and monitoring petroleum hydrocarbons. In: *Proceedings of Petroleum Hydrocarbons and Organic Chemicals in Ground Water: Prevention, Detection and Restoration*. National Water Well Association. Dublin, Ohio. pp. 357-367.

McKee, J.E., F.B. Laverty, and R.M. Hertel. 1972. Gasoline in groundwater. *J. Water Pollut. Cont. Fed*. 44(2):293-302.

McNabb, J.F., and W.J. Dunlap. 1975. Subsurface biological activity in relation to ground-water pollution. *Ground Water*. 13(1):33-44.

Moltyaner, G.L., and R.W.D. Killey. 1988. Twin Lake tracer tests: Transverse dispersion, *Water Resourc. Res*. 24(10):1612-1627.

Molz, F.J., M.A. Widdowson, and L.D. Benefield. 1986. Simulation of microbial growth dynamics coupled to nutrient and oxygen transport in porous media. *Water Resourc. Res*. 22(8):1207-1216.

Mihelcic, J.R., and R.G. Luthy. 1991. Sorption and microbial degradation of naphthalene in soil-water suspensions under denitrification conditions. *Environ. Sci. Tech*. 25(1):169-177.

Odencrantz, J.E., A.J. Valocchi, W. Bae, and B.E. Rittman. 1989. Biodegradation of halogenated solvents by biologically active zones induced by nitrate injection into porous medium flow: II. Computer international modeling. In: *Symposium on Processes Governing Movement and Fate of Contaminants in Subsurface Environment*. IAWPRC. Stanford, California.

Payne, T. B. and P.B. Durgin. 1988. Hydrocarbon vapor concentrations adjacent to tight underground gasoline storage tanks. In: *Proceeding of 2nd Outdoor Action Conf. on Aquifer Restoration, Ground Water Monitoring and Geophysical Methods*. National Water Well Association. Dublin, Ohio. pp. 1173-1188.

Ridgeway, H.F., J. Safarik, D. Phipps, P. Carl, and D. Clark. 1990. Identification and catabolic activity of well-derived gasoline degrading bacteria from a contaminated aquifer. *Appl. Environ. Microbiol.* 56(11):3565-3575.

Rifai, H.S., P.B. Bedient, J.F. Haasbeek, and R.C. Borden. 1989. *BIOPLUME II: Computer Model of Two-Dimensional Contaminant Transport Under the Influence of Oxygen Limited Biodegradation in Ground Water, User's Manual- Version 1.0*. U.S. Environmental Protection Agency. Washington, DC. EPA/600/S8-88/093. 73 p.

Schink, B. 1985. Degradation of unsaturated hydrocarbons by methanogenic enrichment cultures. *FEMS Microbiol. Ecol.* 31:69-77.

Siegel, D.I. 1987. *Geochemical Facies and Mineral Dissolution Bemidji, Minnesota, Research Site*. U.S. Geological Survey Open File Report 87-109. pp. C13-C16.

Smith, R.L. and J.H. Duff. 1988. Denitrification in a sand and gravel aquifer. *Appl. Environ. Microbiol.* 54(5):1071-1078.

Smolenski, W.J., and Suflita, J.M. 1987. Biodegradation of cresol isomers in anoxic aquifers. *Appl. Environ. Microbiol.* 53(4):710-716.

Thomas, J.M., V.R. Gordy, S. Fiorenza, and C.H. Ward. 1990. Biodegradation of BTEX in subsurface materials contaminated with gasoline: Granger, Indiana. *Water Sci. Technol.* 24:(6)53-62.

Troutman, D.E., E.M. Godsy, D.F. Goerlitz, and G.G. Ehrlich. 1984. *Phenolic Contamination in the Sand-and-Gravel Aquifer from a Surface Impoundment of Wood Treatment Wastes, Pensacola, Florida*. U.S. Geological Survey Water Resour. Invest. Rept. 84-4230. 36 p.

U. S. Environmental Protection Agency. 1975. Biological denitrification. Chapter 5. In: *Process Design Manual for Nitrogen Control*. 64 pp.

van Beelen, P., and A.K. Fleuren-Kemila. 1989. Enumeration of anaerobic and oligotrophic bacteria in subsoils and sediments. *J. Contaminant Hydrol.* 4:275-284.

van den Berg, L. 1974. Assessment of methanogenic activity in anaerobic digestion: Apparatus and method. *Biotech. Bioeng.* 16:1459-1469.

Vogel, T.M., and D. Grbic-Galic. 1986. Incorporation of oxygen from water into toluene and benzene during anaerobic fermentative transformation. *Appl. Environ. Microbiol*. 52(1):200-202.

Wilson, B.H., J.T. Wilson, D.H. Kampbell, B.E. Bledsoe, and J.M. Armstrong. 1990. Biotransformation of monoaromatic and chlorinated hydrocarbons at an aviation gasoline spill site. *Geomicrobiology J.* 8:225-240.

Wilson, J.T., J.F. McNabb, J.W. Cochran, T.H. Wang, M.B. Tomson, and P.B. Bedient. 1985. Influence of microbial adaptation on the fate of organic pollutants in ground water. *Environ. Toxicol. Chem.* 4:721-726.

Wilson, J.T., D.H. Kampbell, and J.M. Armstrong. 1993. Natural bioreclamation of alkylbenzenes (BTEX) from a gasoline spill in methanogenic ground water. In: *In Situ and On Site Bioreclamation Conference*. San Diego, California. April 5-8, 1993.

White, D. C., G. A. Smith, M. J. Gehron, J. H. Parker, R. H. Findlay, R. F. Martz and H. L. Fredrickson. 1983. The ground water aquifer microbiota: Biomass, community structure, and nutritional status. *Dev. Ind. Microbiol*. 24:201-211.

Young, L. Y. 1984. Anaerobic degradation of aromatic compounds. In: *Microbial Degradation of Aromatic Compounds*. Ed., D.T. Gibson. Marcel Dekker, Inc. New York. pp. 487-523.

Zehnder, A.J.B.. 1978. Ecology of methane formation. In: *Water Pollution Microbiology, Vol. 2*. Ed., R. Mitchell. John Wiley & Sons, Inc. pp. 349-376.

Zeyer, J., E.P. Kuhn, and R.P. Schwarzenbach. 1986. Rapid microbial mineralization of toluene and 1,3-dimethylbenzene in the absence of molecular oxygen. *Appl. Environ. Microbiol.* 52(4):944-947.

Zeyer J., E.J. Dolfing, and R.P. Schwarzenbach. 1990. Anaerobic degradation of aromatic hydrocarbons. *Biotechnology and Biodegradation*. 33 p.

Zobell, C. E. 1946. Action of microorganisms on hydrocarbons. *Bacteriol. Review*. 10:1-49.

SECTION 10

NATURAL BIOREMEDIATION OF CHLORINATED SOLVENTS

Timothy M. Vogel
Environmental and Water Resources Engineering
Department of Civil and Environmental Engineering
The University of Michigan
Ann Arbor, MI 48109-2125

10.1. SUMMARY

Halogenated solvents, as some of the most mobile constituents of hazardous wastes, can pose a threat to subsurface drinking water supplies. Fifteen years ago, many of these highly chlorinated organic compounds were considered recalcitrant to biological degradation in the environment. Since then, researchers have shown that these compounds undergo chemical reactions with half-lives ranging from days to years. Their transformation products may contain no halogens, as a result of hydrolysis, or still be partially halogenated alkenes, as a result of elimination of hydrogen halide. Elimination reactions, rather than hydrolysis reactions, will predominate with greater halogenation. Microbially-mediated reactions of chlorinated solvents usually involve oxidation or reduction reactions. Oxidation reactions are generally slower with highly halogenated compounds than with compounds containing fewer halogen substituents, while the opposite is true for reduction reactions. Oxidation reactions do not dehalogenate in the first rate-limiting step, but in subsequent steps. Reduction reactions normally include the dehalogenation of these solvents, producing less halogenated homologues. The dechlorination occurs under anaerobic conditions and results in less chlorinated, and often aerobically degradable, products. Engineered systems, or in-situ bioremediation, can effectively employ either aerobic alone or sequential anaerobic/aerobic microbial processes to biodegrade chlorinated solvents. The natural bioremediation of chlorinated solvents depends on the appropriate subsurface environmental conditions. These conditions must promote growth of either anaerobic or aerobic microorganisms and allow for the contact between chlorinated solvent and these microbes. The rate of natural bioremediation may be slow enough that analyses of subsurface chemical constituents would indicate the presence of several chlorinated products of original chlorinated solvents. Enhanced bioremediation requires improving microbial growth conditions, reducing mass transfer limitations, and controlling movement of subsurface chlorinated solvents.

10.2. FUNDAMENTAL PRINCIPLES

Hazardous wastes contain many different classes of compounds, including metals, polyaromatic hydrocarbons (PAH), polychlorinated biphenyls (PCB), aromatic hydrocarbons (e.g., benzene), and halogenated solvents. Large volumes of both chlorinated aliphatic and aromatic hydrocarbons are produced each year for a variety of domestic and commercial purposes (Merian and Zander, 1982; Pearson, 1982). Table 10.1 contains a list of some halogenated solvents and their annual production rates. They have become widely distributed in the environment as a result of discharges from industrial and municipal wastewaters, urban and agricultural runoff, leachates from landfills, and leaking underground tanks and pipes. Sediments beneath some industrial sites contain chlorinated hydrocarbons in excess of 1,000 ppm (Phelps et al., 1990). Several examples of well-documented contaminant plumes in ground water exist both in the United States and Canada (Mackay and Cherry, 1989). Most of these plumes are composed of large quantities (0.4 x 10^9 to 5.7 x 10^9 liters) of water contaminated by chlorinated industrial solvents, such as tetrachloroethylene (PCE), trichloroethylene (TCE), and 1,1,1-trichloroethane (TCA), or aromatic hydrocarbons, such as benzene, toluene, and xylene (Mackay and Cherry, 1989). Such plumes pose problems regarding containment and remediation. Halogenated solvents are relatively mobile in the environment, being both highly volatile and generally less retarded in ground water than many other constituents of hazardous wastes. Due to their relatively high mobility, halogenated solvents can be transported from hazardous waste sites to drinking water wells by ground water. Even a spill of relatively small volume can contaminate millions of gallons of drinking water. A survey conducted by the U.S. Environmental Protection Agency documented that 22% of approximately 466 randomly sampled subsurface drinking water sources contain mixtures of volatile organic chemicals at detectable levels (Symons et al.,1975; Westrick et al., 1984). Some halogenated solvents potentially pose significant human health hazards, which is in part reflected in the drinking water maximum contaminant levels (MCLs) set by the U.S. Environmental Protection Agency. These MCLs range from near 1 microgram per liter (µg/l) for compounds such as vinyl chloride to 200 µg/l for TCA (Table 10.1). The ultimate fate of these halogenated aliphatic compounds in ground-water supplies is often controlled by their chemical and biological reactivity. Although chlorinated organic solvents have been released into the environment for decades, they have only come under intense international scrutiny in the last 10 years. Investigations of the fate of these compounds in the environment, including volatilization into the atmosphere, sorption onto sediments, bioaccumulation and concentration in aquatic and terrestrial organisms, and dissolution in surface and ground waters have led to increased understanding of the movement of these compounds. Yet, none of these processes actually degrades these compounds.

TABLE 10.1. PRODUCTION, PROPOSED MAXIMUM CONTAMINANT LEVELS, AND TOXICITY
RATINGS OF COMMON HALOGENATED ALIPHATIC COMPOUNDS[a]

Compound	Production[b] (million lb/yr)	MCL[c] (mg/l or ppm)	Carcino-genicity[d]
Trihalomethanes	- -	100	- -
Vinyl chloride	7000	1	1
1,1-Dichloroethylene	200	7	3
trans-1,2 Dichloroethylene	<0.001	--	--
Trichloroethylene	200	5	3
Tetrachloroethylene	550	--	--
1,1-Dichloroethane	<0.001	--	--
1,2-Dichloroethane	12,000	5	2
1,1,1-Trichloroethane	600	200	3

a Vogel et al., 1987.
b Federal Register, 1985.
c Maximum contaminant level, Van Nostrand Reinhold Co., 1984.
d Carcinogenicity: 1 = chemical is carcinogenic; 2 = chemical probably is carcinogenic;
   3 = chemical cannot be classified.

Indeed, these compounds were considered resistant to chemical and biological degradation, although medical studies had shown that some chlorinated compounds are metabolized by rat liver cells (Anders, 1983). One reason for the apparent recalcitrance of highly chlorinated compounds is their oxidized nature. The reactivities of chlorinated solvents are controlled mainly by the degree of halogenation. Chlorine substituents tend to make the chlorinated solvent fairly oxidized, unlike most hydrocarbons composed of only carbon and hydrogen. The specific chemistry of these solvents controls the types of reactions they undergo. Due to the variety of the reactions that halogenated solvents undergo, a diverse range of transformation products may be found at contaminated sites. The products often have fewer chlorine substituents than the original compound, and may be more or less toxic. In some cases, the halogenated solvents are completely mineralized to carbon dioxide (Vogel et al., 1987). An understanding of the changes that chlorinated compounds are subject to in the environment provides the basis for understanding their natural bioremediation and for designing of remediation systems. Conditions can be induced that stimulate the transformations of hazardous compounds to the least harmful products possible. A crucial point in the complete destruction of chlorinated hydrocarbons, as will be discussed, is the removal of chlorine substituents from the molecule.

10.3. CHEMICAL REACTIONS

Halogenated solvents generally undergo either or both substitution and dehydrohalogenation reactions in water (Vogel et al., 1987). Substitution reactions of halogenated solvents in water (hydrolysis) involves the replacement of a halogen substituent with a hydroxy (-OH) group, forming an alcohol. For example, chloroethane is hydrolyzed to ethanol (Vogel and McCarty, 1987a). If the halogenated solvent has more than one halogen, further substitution reactions can occur. For example, TCA is partially hydrolyzed through a series of substitution reactions to acetic acid (Mabey et al., 1983); and 1,2-dibromoethane (EDB) is partially hydrolyzed to ethylene glycol (Weintraub et al., 1986). Dehydrohalogenation reactions of halogenated solvents in water usually involve the elimination of hydrogen halide from an alkane and the formation of an alkene. Some halogenated solvents undergo both hydrolysis and dehydrohalogenation in water. For example, the two compounds described to undergo hydrolysis above, TCA and EDB, also form 1,1-dichloroethylene (1,1-DCE) (Vogel and McCarty, 1987b) and vinyl bromide (bromoethylene) (Vogel and Reinhard, 1986), respectively, as a result of dehydrohalogenation. The likelihood that a halogenated solvent will undergo either hydrolysis or dehalogenation depends in part on the number of halogen substituents. More halogen substituents on a compound tend to increase the chance of dehydrohalogenation reactions occurring. The opposite is true for hydrolysis reactions. Bromine substituents are generally more reactive than chlorine substituents. So, less of these substituents than chlorine substituents are needed to cause the halogenated solvent to undergo dehydrohalogenation. Location of the halogen substituents on the carbon skeleton also has some effect on both type and rate of reaction.

The rates of substitution and dehydrohalogenation reactions are also dependent on the degree of halogenation. Substitution reactions generally decrease in rate with increasing halogenation. Monohalogenated alkanes have half-lives at 25°C on the order of days to months. Polychlorinated alkanes have half-lives that range up to thousands of years for carbon tetrachloride. Dehydrohalogenation rates for halogenated solvents increase with increasing halogenation. Unfortunately, many reported environmentally significant chlorinated solvent half-lives are the result of extrapolation from experiments performed at elevated temperatures (Mabey and Mill, 1978). Some chlorinated solvent reaction rates are so slow as to make experiments run at environmental temperatures impractical. However, in order to accurately extrapolate rate values from elevated temperatures, experiments need to be conducted at several different temperatures. The data can be extrapolated using the Arrhenius equation in a manner that includes statistical evaluation (Vogel and Reinhard, 1986). The resulting range of values for chlorinated solvent half-lives can be used to estimate lifetimes of chemicals in ground water. However, other processes such as sorption will influence these estimates. Typically, the activation energies for chlorinated solvent hydrolysis and dehydrohalogenation reactions are approximately 100 KJoules/mole, which results in a factor of 3.5 change in reaction rate and half-life with each 10°C change in temperature. So, values listed at 25°C, such as three years for TCA would be 10.5 years at 15°C. Beyond the impact of the initial chemicals themselves, the approximate half-lives and the potential products of chlorinated solvent chemical reactions in water tend to suggest that compounds that undergo dehydrohalogenation can cause the most significant health hazard. In addition, natural bioremediation can also act on both the original compound and its chemical (or biological) products, leading in some cases to a mixture of chlorinated compounds in the subsurface.

10.4. MICROBIOLOGICAL REACTIONS

Chlorinated solvents can undergo both substitution and elimination reactions similar to those described above, yet mediated by microorganisms. However, the most common reactions are those involving the transfer of electrons to or away from the chlorinated solvent. These reactions, oxidation for removal of electrons from chlorinated solvents and reductions for addition of electrons to chlorinated solvents, are dependent to some extent on the degree of chlorination of the chlorinated solvent and upon the redox conditions of the microorganisms (Vogel et al., 1987). The more highly chlorinated solvents are more highly oxidized and, therefore, are more likely to undergo reduction reactions. The less chlorinated solvents are less oxidized and are more likely to undergo oxidation reactions. Redox conditions vary from the most oxidative (+800 mv), where oxygen is present and aerobic microbes grow, to the most reduced, where neither oxygen, nitrate nor sulfate exists, and methanogens grow (~ -350 mv). Hence, aerobic conditions should be more suitable to the oxidation of less chlorinated solvents and methanogenic conditions should be more suitable to the reduction of highly chlorinated solvents. Oxidations of chlorinated solvents usually involve the addition of oxygen without the removal of halogen in the first and rate-limiting step (Vogel et al., 1987). Subsequent release of chlorine substituents might not actually be associated with oxidation reactions, but with substitution reactions.

10.4.1. Aerobic

Most research to date has described the microbial oxidations of mono- or dihalogenated aliphatic compounds. The major exception to this is the work done on the oxidation of trichloroethylene (TCE). Several different microbes or microbial enrichments have been shown to be capable of TCE oxidation (Fogel et al., 1986; Nelson et al., 1986; Little et al., 1988) and chloroform oxidation (Strand and Shippert, 1986). Apparently, the ease of oxidation increases with decreasing number of halogens. Hence, dichloroethylene would be oxidized faster than TCE. Unfortunately, due to the nature of contaminant release in the environment, mass balances are difficult to achieve, and no strong evidence for the oxidation of halogenated solvents has been derived from actual hazardous waste sites.

Highly chlorinated organic compounds are much more oxidized than many natural organics. As such, these compounds do not provide much energy upon further oxidation in aerobic environments. Most aerobic biodegradation processes start with a step that involves the insertion of oxygen into a bond on the molecule. Due to the electrophilic nature of that oxygen insertion, other electrophilic substituents (e.g., chlorine) hinder the reaction. Hence, the observation that increasing chlorination within a homologous series often leads to a decrease in aerobic (oxidative) biodegradation (Vogel et al., 1987).

Studies of the aerobic biodegradation of chlorinated compounds have illustrated several major pathways of oxidation. These pathways resemble those for the nonchlorinated homologues. For example, the oxidation of chlorinated ethylenes involves the formation of a chlorinated epoxide similar to that for ethylene:

Example:

$$
\begin{array}{ccc}
\mathrm{Cl} \quad\quad \mathrm{H} & & \mathrm{Cl}\ \ \mathrm{O}\ \ \mathrm{H} \\
\diagdown\quad\diagup & & \diagdown\diagup\diagdown\diagup \\
\mathrm{C = C} & \longrightarrow & \mathrm{C - C} \\
\diagup\quad\diagdown & & \diagup\quad\ \diagdown \\
\mathrm{H} \quad\quad \mathrm{Cl} & & \mathrm{H} \quad\ \ \mathrm{Cl}
\end{array} \qquad (1)
$$

The epoxide degrades rapidly in water. In both of these cases, the microbe that degrades these compounds might require a natural nonchlorinated compound for growth and energy. The enzymes produced for degradation of that "normal" substrate are also capable of degrading the pollutant (cometabolism). The possibility that these microbes would adapt to the use of chlorinated compounds as sources of energy and carbon exists, but might have limited engineering applications. Selective pressure in natural environments will not be great if pollutant concentrations are relatively low from a microbial adaptation point of view, even if these concentrations are high from a regulatory point of view.

In other aerobic degradation of lightly chlorinated compounds, microbes have been shown to grow on the pollutant when it exists in sufficiently high concentration. Most of these compounds are mono- or dichlorinated organics. A common reaction is the microbially mediated substitution reaction where a hydroxyl group replaces a chlorine (Brunner et al., 1980).

Example:

$$
\begin{array}{ccc}
& & \mathrm{H} \\
& & | \\
\mathrm{Cl}\ \ \mathrm{Cl} & & \mathrm{O}\ \ \ \mathrm{Cl} \\
|\ \ \ | & & |\ \ \ \ | \\
\mathrm{H - C\ - C - H} & \longrightarrow & \mathrm{H - C\ - C - H} \\
|\ \ \ | & & |\ \ \ \ | \\
\mathrm{H}\ \ \mathrm{H} & & \mathrm{H}\ \ \mathrm{H}
\end{array} \qquad (2)
$$

After which, the compound is further oxidized and the metabolites enter the anabolic and catabolic pathways of the microbe.

Although progress has been made in culturing aerobes capable of degrading organic compounds with higher and higher degrees of chlorination, many highly chlorinated compounds remain resistant due to their highly oxidized state. Examples of aerobically recalcitrant chlorinated organics include tetrachloroethylene, which has not been observed to undergo epoxidation, and hexachlorobenzene, which have all carbons occupied with chlorine substituents, allowing no site for hydroxylation. These highly chlorinated organic compounds are not, however, resistant to anaerobic biodegradation (Vogel and McCarty, 1985; Gibson and Suflita, 1986; Tiedje et al., 1987; Vogel and McCarty, 1987a; Vogel, 1988; Freedman and Gossett, 1989; Bagley and Gossett, 1990; Nies and Vogel, 1990; Bhatnagar and Fathepure, 1991).

10.4.2. Anaerobic

Under different anaerobic conditions, both in laboratory studies and in the environment, highly chlorinated organics, such as tetrachloroethylene (Vogel and McCarty, 1985), hexachlorobenzene (Gibson and Suflita, 1986), and polychlorinated biphenyls (Nies and Vogel, 1990), have been shown to undergo reductive dechlorination. Reductions of chlorinated solvents normally involve the removal of a chlorine substituent and either its replacement with a hydrogen or removal of a second chlorine substituent from alkanes and formation of an alkene. The first mechanism, commonly called reductive dechlorination, can occur with both alkanes and alkenes. Reductive dechlorination has been described for the sequence of ethylenes from tetrachloroethylene to vinyl chloride (Vogel and McCarty, 1985) (Figure 10.1) and to ethylene (Freedman and Gossett, 1989), and for TCA to chloroethane (Vogel and McCarty, 1987a) (Figure 10.2) under methanogenic conditions in laboratory studies. In the case of TCA, potential products are complicated by the chemical reactions (denoted by A) that co-occur with biological reactions. Relative rate studies on the reductive dechlorination of various chlorinated ethanes and ethenes have shown a general decrease in rate with decreasing number of chlorine substituents, opposite to the trend shown for oxidation reactions under aerobic conditions. The relative rates of reduction under methanogenic conditions have been quantified in two cases (Table 10.2) (Bouwer and McCarty, 1988; Vogel, 1988). From these data and that for the chemical reactivity described above, the disappearance of an initial chlorinated solvent and the appearance of its products under favorable anaerobic conditions might be derived, as will be discussed later.

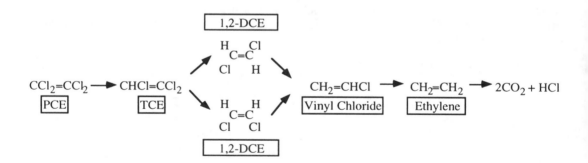

Figure 10.1.   PCE anaerobic transformations.

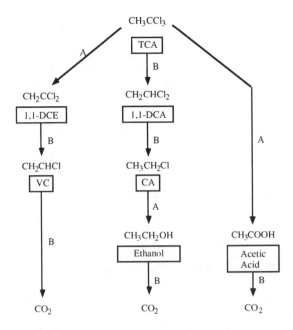

Figure 10.2   Abiotic (A) and biotic (B) transformations of 1,1,1-trichloroethane.

TABLE 10.2.   RELATIVE RATES OF DEGRADATION BY METHANOGENIC CULTURES

Compound	Suspended Growth[a]	Attached Growth[b]
Carbon Tetrachloride	--	3.3
Chloroform	--	1.3
Trichloroethylene	1.04	--
Acetate	1.00	1.00
1,1-Dichloroethylene	0.55	--
1,1,1-Trichloroethane	0.45	0.73
1,1,2-Trichloroethane	0.43	--
1,2-Dichloroethane	0.05	--
1,1-Dichloroethane	0.04	--
Chloroethane	0.02	--
Vinyl Chloride	0.005	--

[a] Vogel, 1988.
[b] Bouwer and McCarty, 1988.

Possibly the observed reductive dechlorination of chlorinated solvents results from the reduction-oxidation reaction between the highly oxidized chlorinated compound and reduced compounds in the microbe. Bacteria contain a number of metal-organic compounds, which are involved in electron exchange reactions. Many of these metal-organic compounds, when in their reduced form, have been shown to dechlorinate chlorinated organics (Gantzer and Wackett, 1991; Assaf-Anid et al., 1992). The metal-organic compound becomes oxidized as a result of this reaction, but might be reduced again in the normal metabolic processes of the cell. In this case, the ultimate source of electrons would be the organic substrate (e.g., acetate) that the anaerobe is using for energy and carbon. The thermodynamics are very favorable for the oxidation of metal-organics and concomitant reduction of chlorinated compounds. The fortuitous nature of this reaction and the requirement for an ultimate external electron donor have significant implications for bioremediation of these compounds.

On the other hand, anaerobic microorganisms might adapt to the use of a chlorinated compound as an alternative terminal electron acceptor. If this occurs, the chlorinated compound degradation would be directly tied to energy production in the cell. Although this type of anaerobic dechlorination mechanism has been observed, it might not be very common, due to typically low concentrations of dissolved chlorinated organics in the environment.

10.5. PREDICTIONS OF PRODUCT DISTRIBUTION

The natural biodegradation of chlorinated solvents discussed above leads to predictions of the natural bioremediation of these compounds. The next section describes the outcome of natural anaerobic reductive dechlorination of the chlorinated solvents, TCE and TCA, based on relative rates (Table 10.2). One of the major unknown aspects regarding the microbial transformations of halogenated solvents in ground water is the number and activity of appropriate microbes. Research is needed for understanding the diversity of occurrence of these microorganisms. The simplest case is the addition of PCE to an anaerobic aquifer. Assuming a particular microbial activity, PCE is transformed to TCE, and then to 1,2-DCE, which is subsequently transformed to vinyl chloride (VC) as described previously. A simple model using relative rates determined in lab studies (Vogel, 1988, Table 10.2) can show the distribution of products over time or distance in an active anaerobic subsurface zone (Figure 10.3). A more active microbial population would increase the rate of transformations. For the case where TCA has contaminated an anaerobic aquifer, the transformations are complicated by the potential simultaneous chemical and microbial reactions. The simplest TCA case is where no microbial activity exists at all. In this case, only acetic acid and 1,1-DCE are formed due to hydrolysis and dehydrohalogenation, respectively (Figure 10.4). When some methanogenic activity exists, TCA is partially reduced to 1,1-dichloroethane (DCA), which is subsequently reduced to chloroethane, which can be, in some cases, mineralized to carbon dioxide (Figure 10.5). Furthermore, the 1,1-DCE from the dehydrohalogenation of TCA is reduced to vinyl chloride (VC) (Figure 10.5). Acetic acid is also microbially mineralized partially to carbon dioxide by methanogens. If the microbial activity increases, then the pathways are dominated by the microbial reactions, not the chemical reactions. In this case, DCA, VC, and carbon dioxide (CO_2) predominate after time (Figure 10.6). The most complicated case presented here is when both TCE and TCA contaminate an anaerobic aquifer. When appropriate microbial activity occurs, dichloroethylene isomers (1,2-DCE from TCE and 1,1-DCE from TCA) are reduced to VC (Figure 10.7). TCA is reduced to DCA. Note that at one point in time or space, the major products are DCA and the DCE isomers. Remember that if the ground water should carry the mixture of chlorinated solvents out of an active microbial (methanogenic) area, then most of the reductive dechlorination reactions would stop. The ratios of chemical compounds would not change, except due to the influence of other processes, such as aerobic degradation.

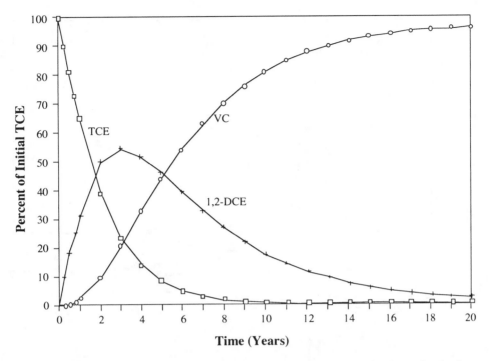

Figure 10.3. Reductive dechlorination of trichloroethylene (TCE) under hypothesized anaerobic field or laboratory conditions.

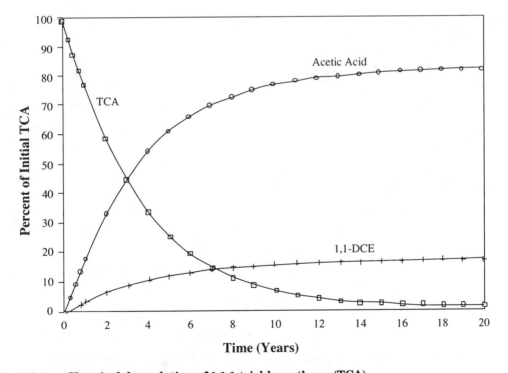

Figure 10.4. Chemical degradation of 1,1,1-trichloroethane (TCA).

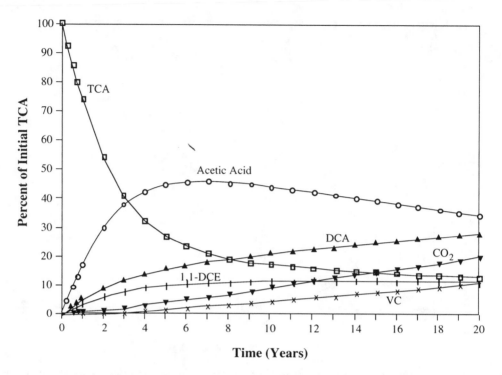

Figure 10.5.   Chemical and microbial degradation of TCA (lower microbial activity).

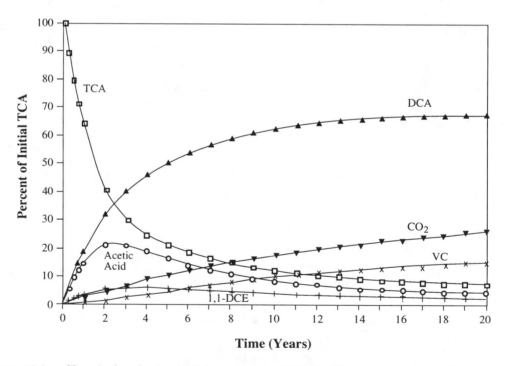

Figure 10.6.   Chemical and microbial degradation of TCA (higher microbial activity).

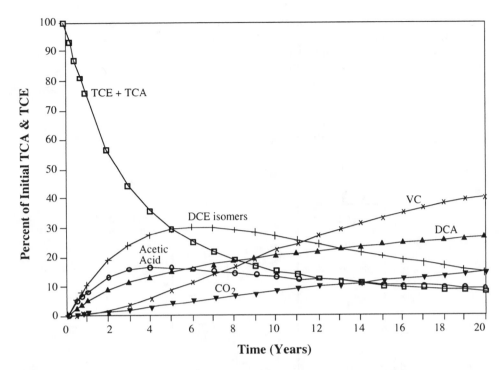

Figure 10.7.   Chemical and microbial degradation of both TCE and TCA.

10.6.   RATIONALE FOR TECHNOLOGY

Many examples of the transformation of halogenated compounds under anaerobic conditions by reductive dechlorination (Vogel and McCarty, 1985; Vogel and McCarty, 1987a; Vogel et al., 1987; Freedman and Gossett, 1989; Bagley and Gossett, 1990) have been published, supporting the effectiveness of this first step for chlorinated solvent degradation.   Reductive dechlorination as described above is relatively rapid for chemicals with a higher number of chlorine substituents, such as highly chlorinated PCBs, hexachlorobenzene (HCB), perchloroethylene (PCE), trichlorethylene (TCE), carbon tetrachloride (CT), chloroform (CF) and 1,1,1-trichloroethane (TCA) when compared with their less chlorinated homologues (Tiedje et al., 1987; Vogel and McCarty, 1987a; Vogel et al., 1987; Bouwer and Wright, 1988; Fathepure et al., 1988).   Upon reduction, these polychlorinated compounds lose chlorine, and the resulting products are usually more susceptible to hydrolytic and oxidative processes and less susceptible to further reduction.   These lower chlorinated compounds have been shown to be successfully degraded by aerobic bacteria (Kuhn et al., 1985; de Bont et al., 1986; Schraa et al., 1986; Strand and Shippert, 1986; Spain and Nishino, 1987; van der Meer et al, 1987; Vogel et al, 1987; Henson et al., 1988).   Therefore, the anaerobic/aerobic sequential biodegradation of highly chlorinated compounds by indigenous microbes could occur and should be encouraged.

In order for compounds to undergo natural anaerobic/aerobic sequential environmental conditions, compounds would have to diffuse or flow from anaerobic zones to aerobic zones.   This could occur near sites that contain easily degradable reduced organics, thus consuming oxygen near the source of contamination.

This scheme requires the establishment of anaerobic conditions, followed by aerobic conditions, which is not the normal ecological trend. This sequence can, in some cases, be physically modeled spatially in a flow-through system (e.g., Figure 10.8) or in chronological order with the same material.

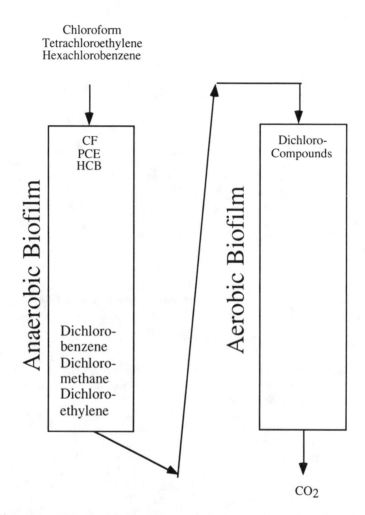

Figure 10.8.   **Schematic illustrating the reductive dechlorination of polychlorinated compounds in an anaerobic biofilm and subsequent mineralization of the products of anaerobic treatment in an aerobic biofilm.**

In most systems described, the anaerobic microbial conditions resulted in dechlorination of the highly chlorinated organic compounds, although none of the compounds were completely dechlorinated, with the exception of vinyl chloride going to ethylene (Freedman and Gossett, 1989). Typically, mono-, di-, and trichlorinated compounds remained after the anaerobic phase was complete (Table 10.3). In a physical model, relative amounts of mono-, di-, and trichlorinated dechlorination products varied depending on the organic substrate and nutrients fed to the system and the hydraulic residence time, among other parameters (Fathepure and Vogel, 1991).

TABLE 10.3. PRODUCTS OF ANAEROBIC DECHLORINATION

Initial Compound		Major Products
Hexachlorobenzene	→	di- and trichlorobenzene
Tetrachloroethylene	→	1,2-dichloroethylenes
Trichloromethane	→	dichloromethane
Hexachlorobiphenyl	→	trichlorobiphenyl

Aerobic degradation of mono- and dichlorinated organic compounds was fairly rapid in several systems tested (Brunner et al., 1980; Fogel et al., 1986; Nelson et al., 1986; Henson et al., 1988; Walton and Anderson, 1990; Fathepure and Vogel, 1991). When carbon-14 radiolabelled compounds were used to track the carbon, considerable amounts of carbon-14 labelled carbon dioxide were produced (Table 10.4) (Walton and Anderson, 1990; Fathepure and Vogel, 1991). The most recalcitrant chlorinated solvents of those that underwent aerobic degradation were the trichlorinated compounds. In some cases, these compounds are (as were the mono- and dichlorinated compounds) the result of reductive dechlorination of more chlorinated compounds under anaerobic conditions.

TABLE 10.4. PRODUCTS OF AEROBIC DEGRADATION[a]

Initial Compound	Product(s)
Dichlorobenzene	CO_2
Dichloroethylenes	CO_2
Dichloromethane	CO_2

[a] Fathepure and Vogel, 1991.

Anaerobic reductive dechlorination in these studies was dependent on organic substrate and nutrients. The solution shown in Table 10.5 represents an example of nutrients for anaerobic bacteria. In many subsurface environments, these nutrients may already exist dissolved in ground water. Another critical addition is the reduced organic that will supply energy and carbon for the growth of anaerobes. Further, it will be the ultimate source of electrons for the reductive dechlorination of the chlorinated organic compounds. In addition, its degradation by aerobes will also deplete oxygen, keeping the region anaerobic. Products (such as methane) of the anaerobic degradation of these substrates might provide substrates for aerobic microbes later.

TABLE 10.5.   PROPOSED NUTRIENTS FOR BIOREMEDIATION

Compound	Concentration (milligrams per liter)
$(NH_4)_2HPO_4$	80
NH_4Cl	1,000
K_2HPO_4	200
$NaCl$	10
$CaCl_2$	10
$MgCl_2$	50
$CoCl_2, 6H_2O$	1.5
$CuCl_2, 2H_2O$	0.2
Na_2MoO_4, H_2O	0.23
$ZnCl_2$	0.19
$NiSO_4, 6H_2O$	0.2
$FeSO_4, 7H_2O$	1.0
$AlCl_3, 6H_2O$	0.4
H_3BO_3	0.38

Several important implications can be derived from these laboratory studies. The sequential anaerobic/aerobic biodegradation of chlorinated organic compounds might be a viable treatment technology, either naturally or induced. The anaerobic phase requires the induction of active anaerobic metabolism (e.g., methane production) by a consortium of anaerobes, not necessarily one specific microbe. This anaerobic consortium must be supported by nutrients and organic substrate(s). If these compounds already exist in solution, they need not be added. Due to the apparent lack of dependence of the anaerobes on the chlorinated compounds for growth in many cases, the concentration of these chlorinated solvents can theoretically be reduced to zero.

For the aerobic phase, oxygen (possibly in the form of hydrogen peroxide) must be added or mixed in naturally in order to oxidize the anaerobic conditions. Viable aerobes capable of degrading chlorinated compounds must either be present or be added. In cases where the compounds are sufficiently high in concentration, these aerobes will use the chlorinated compound as a source of energy and carbon. Hence, the theoretical minimum concentration achievable would be the lowest concentration capable of supporting the microbial community. Possibly, other substrates that will not out-compete the chlorinated compound, but will aid in supporting the microbial community, can be added or already exist. Additional substrates are required when the chlorinated compounds are too low in concentration to induce enzymes or support aerobic microbial growth. The sequential anaerobic/aerobic biodegradation of chlorinated organic compounds in laboratory studies provides a sound, but not complete, basis for field application testing.

Many cases of natural anaerobic reductive dechlorination of chlorinated solvents in ground waters have been observed by consulting engineers, who question the apparent production of less chlorinated products. In some cases, these products seem to disappear once they reach aerobic zones. The research site at St. Joseph, Michigan, might provide evidence of this natural combination of anaerobic (McCarty and Wilson, 1992) and aerobic processes.   The plume contains high concentrations of methane and relatively low concentrations of chlorinated solvents at the time it discharges to surface water. Natural cooxidation of the chlorinated compounds during aerobic metabolism of the methane in the oxygenated benthic sediments at the interface between the plume and surface water is likely. However, the phenomenon has not been carefully documented.

To date, no organism has been found which can use the anaerobic metabolism of halogenated solvents as the sole source of carbon and energy. Instead, microorganisms generally require a primary source of carbon and energy. Although the reduction of a halogenated solvent (the secondary substrate) is usually energetically favorable at standard state, the organism may or may not benefit from this reduction due to inability to control energy enzymatically or due to low concentrations leading to relatively little available energy. The relationship may be fortuitous. A proposed mechanism for the reductive dechlorination of halogenated solvents by methanogens, the group of organisms most commonly cited as being responsible for reductions, involves the transfer of electrons through the cell membrane during the anaerobic metabolism of primary substrate (Zeikus et al., 1985). In this case, the halogenated solvent may divert some of these electrons and use them for dechlorination. Hence, it is conceivable that microbes might use the chlorinated solvent as an electron acceptor.

10.7.   PRACTICAL IMPLICATIONS

The implication of a microbial degradation process that is dependent on the primary substrate concentration, but not on the halogenated solvent concentration, is that there is no lower limit to the final concentration of the halogenated solvent. If the chlorinated compound served as a primary substrate, then fewer organisms would survive when its supply became low, and further dechlorination would cease (McCarty, 1984). As a secondary substrate, the halogenated solvent can continue to be reduced by a large, healthy population of bacteria grown on the primary substrate until the halogenated solvent has been completely degraded.

Actual contamination is often a mixture of complex chemicals. Therefore, future technologies aimed at bioremediation of halogenated solvents should achieve complete destruction of all hazardous chemicals.   Based on metabolic and kinetic limitations of anaerobic and aerobic bacteria, a two-stage biological process consisting of an initial anaerobic dechlorination of highly chlorinated chemicals followed by aerobic degradation of the partially dechlorinated metabolites may effectively treat wastes containing complex mixtures of chlorinated hydrocarbons.

In a given environment, any or all of the types of reactions discussed above may occur. The conditions present, as well as the structure of the chlorinated compound, dictate to a large extent the transformations that will predominate, and the expected products. The only chemical reactions that have a significant effect on degradation products are the hydrolysis of monochlorinated solvents and the dehydrohalogenation of chlorinated polyalkanes. However, biological reactions can achieve rates with half-lives as low as a few days, and may be significantly influenced by controlling environmental conditions.

An important factor influencing biological degradation is whether the necessary organisms are present. This should be determined before a full-scale remediation scheme is begun by sampling the aquifer material in the area of the contaminated ground water to be treated, and running laboratory-scale treatability studies.

Microbial substrates and electron acceptors are the next factors to be considered. In a given location, the biological transformations will be oxidation reactions, mediated by aerobic organisms, as long as oxygen is present. Once oxygen has been depleted, alternative electron acceptors such as nitrate and sulfate will be used. Finally, anaerobic (even methanogenic) microorganisms will dominate and reduction reactions will take place. The types of reactions that actually occur, however, may be dictated by the type and amounts of substrates added to the ground water. Thus degradation conditions can occur or be imposed according to the products desired.

In remediation of ground water, the choice of a treatment method should be based on which type of reaction will be the most rapid and whether the expected products are less hazardous than the original halogenated solvent. For example, reduction of PCE leads to the production of vinyl chloride, which is a known carcinogen. However, the above discussion suggests an efficient means of converting halogenated solvents to nonhazardous compounds: sequencing anaerobic and aerobic treatments. PCE would be reduced to TCE and DCE under anaerobic conditions, then an aerobic environment would be provided in which the TCE and DCE would be oxidized to carbon dioxide. Overall degradation rates would be maximized, and no vinyl chloride would be produced (Fathepure and Vogel, 1991).

10.8. SPECIAL REQUIREMENTS FOR SITE CHARACTERIZATION

Considerable site characterization of parameters directly related to in-situ biological activity, in addition to other site characteristics, is required (Table 10.6). As natural bioremediation of chlorinated compounds is controlled in part by the natural redox of the ground water, sites amenable to anaerobic reductive dechlorination require information regarding the ability of indigenous microbes to undergo anaerobioses. An example of a specific characteristic is the dissolved oxygen concentration in the ground water at the site. This information is critical to understanding which type of microbial community is dominant. Clearly, measurement of chemical parameters directly related to microbial activity and not just analyses of priority pollutants is critical for evaluating the likely success of natural bioremediation. For example, as listed in Table 10.6, pH, temperature, ionic strength, presence or absence of heavy metals and of potential electron acceptors all play important roles in determining the type and extent of microbial activity. Laboratory tests that evaluate microbial activity and potential toxicity for a given site will aid in determining potential for natural bioremediation. These methods might measure C-14 carbon dioxide or methane production or, in the case of anaerobic conditions, the production of dechlorinated products.

10.9. FAVORABLE SITE CHARACTERISTICS

Since highly chlorinated solvents (e.g., tetrachloroethylene: PCE) do not appear to undergo aerobic degradation, the degree of chlorination of the solvent is critical for differentiating between whether aerobic (less chlorinated solvents) or anaerobic conditions will be effective. The degree of chlorination will also, to some extent, control the sorption onto organic matter in the aquifer and, thus, the retardation of the solvent through the aquifer or soil. The relative retention of the solvent affects the practicality of potential natural bioremediation, as will be discussed below.

TABLE 10.6. SOME INFORMATION NEEDED FOR PREDICTION OF ORGANIC CONTAMINANT MOVEMENT AND TRANSFORMATION IN GROUND WATER[a]

Hydraulic	*Contaminant Source*	*Wells*	*Hydrogeologic Environment*
	location amount rate of release	location amount depth pump rates	extent of aquifer and aquitard characteristics of aquifer hydraulic gradient ground-water flow rate
Sorption	*Distribution Coefficient*	*Characteristics of the Aquifer Solid*	*Contaminant Characteristics*

	characteristic of
	   concentration
		organic carbon content
		clay content
			octanol/water partition
			   coefficient
			solubility

Chemical	*Ground-water Characteristics*	*Aquifer Characteristics*	*Contaminant Characteristics*
	ionic strength pH temperature NO_3^-, $SO_4^=$, O_2 toxicants	potential catalysts: metals, clays	potential products concentration
Biological	*Ground-water Characteristics*	*Aquifer Characteristics*	*Contaminant Characteristics*

	ionic strength
	pH
	temperature
	nutrients
	   substrate
	   O_2, NO_3^-, $SO_4^=$
	   macro (P, S, N)
	   trace
	organism
	   concentration
	   distribution
	   type
		grain size
		active bacteria -
		   number
		Monod rate - constants
			potential products
			toxicity
			concentration

[a] Vogel, 1988.

The characteristics of the contaminated site also affect the relative mobility of the chlorinated solvents and the potential for natural bioremediation. Hydrologic and geologic conditions that allow for microbes, chlorinated solvents, and necessary nutrients to occur in the same locations at the site will improve the probability of success. The strong sorption of solvents on organic-rich aquifer or soil material will make these compounds less available to microbial activity and, thereby, possibly slow rates. On the other hand, porous subsurface materials, such as sands, are better for ground-water flow but often have less associated microbial activity than organic-rich biodegradation materials and might not be conducive to anaerobic conditions.

The interplay between environmental conditions in the subsurface, proposed remedial actions, and regulatory requirements renders the natural bioremediation of halogenated solvents difficult to pursue. Yet, the possibility for implementing aerobic or anaerobic/aerobic sequential systems successfully are enhanced by relatively porous subsurface material, no heavy metal toxicity, easy access to potential nutrients, ability to reinject recovered contaminated ground water, and hydrogeologic control over the contaminated plume.

10.10. UNFAVORABLE SITE CHARACTERISTICS

Site characteristics that would make the natural bioremediation of halogenated solvents difficult are those that hinder the initiation of appropriate microbial conditions and activities, and those characteristics that prevent the commingling of pollutant with active microbes. As an example, although heavily chlorinated solvents tend to sorb more strongly than less chlorinated solvents, they still undergo reductive dechlorination more rapidly than the less chlorinated solvents under active methanogenesis. For lightly chlorinated solvents, sorption is less important. They tend to undergo degradation under aerobic versus anaerobic conditions (Figure 10.9). Toxic heavy metals, high sulfide concentrations, and lack of appropriate nutrients are chemical characteristics that will negatively affect the natural bioremediation of chlorinated solvents. Hydrologic and geologic characteristics that would not benefit the natural bioremediation of chlorinated solvents include fractured rock systems where small microbial populations exist. Increasing these populations can be difficult, although development of subsurface biofilms with continuous recycling of ground water with appropriate nutrients added might be effective. As mentioned above, physical systems or regulations that prevent mixing of chlorinated solvents and nutrients will hinder or prevent natural bioremediation of chlorinated solvents.

Given the potential difficulties and lack of information regarding site characteristics, the natural bioremediation of chlorinated solvents, even highly chlorinated compounds such as PCE, is still likely at many sites. Indeed, throughout the USA, sites that were initially contaminated with only PCE, TCE, or TCA have shown active dechlorination patterns near the center of contaminated plumes forming the reductive dechlorination products. In some cases, these products appear to be somewhat degraded, as they migrate into aerobic zones. The limitations on these unaided natural processes are currently unknown. Several possibilities exist, such as low microbial activity, for which nutrient addition might be increased.

As mentioned earlier, the extent of degradation in using this approach can theoretically be complete. Since, in most cases, the microbes are not utilizing the chlorinated solvents as food or substrate, microbes could degrade the last molecule, assuming contact exists. The microbes need to be supported on the appropriate substrates and nutrients and be given sufficient opportunity for contact with the chlorinated solvent. Cometabolism defined in this way provides tremendous potential for eventually degrading pollutants to levels below detection.

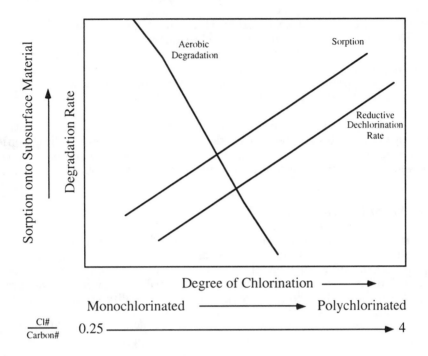

Figure 10.9. **Relationships between degree of chlorination and anaerobic reductive dechlorination, aerobic degradation and sorption onto subsurface material.**

One critical aspect of the natural bioremediation of chlorinated solvents is the rate at which they are degraded. As was mentioned before, the rate is dependent on which types of chlorinated solvent and microbial community interact. Reductive dechlorination rates (estimated as first-order constants or k/ks in Monod terms) suggest that the dechlorination of highly chlorinated compounds might occur in active anaerobic conditions within one year (Vogel, 1988). This rapid microbial dechlorination rate would undoubtedly be controlled by other factors such as sorption/desorption, availability of nutrients, temperature, etc.

A major difficulty with this approach is the lack of predictability of the activity of in-situ microbial communities. In laboratory studies, mainly unreported, some active anaerobic communities have little ability to dechlorinate these solvents. However, in other studies, anaerobic conditions induced in soil and aquifer material samples exhibited reductive dechlorination. Our lack of understanding of the important members of anaerobic microbial communities for reductive dechlorination of chlorinated solvents lends uncertainty to this technology. For aerobic conditions, research has also shown that not all aerobes are active toward lightly chlorinated solvents. An example of this was the lack of TCE-oxidizing ability in all microbes that oxidized aromatic compounds (Nelson et al., 1987; Nelson et al., 1988). Two remedies exist: the first would involve changing the subsurface environmental conditions to activate and grow the appropriate microbes; the second is the addition of selected microbes to the subsurface.

10.11. COST EVALUATIONS

The difficulties in assessing cost and factors that affect costs include the lack of demonstration sites of these processes. Natural unaided bioremediation of chlorinated solvents imposes little or no costs other than the time for the natural processes to proceed. The facilitated natural bioremediation requires control of the subsurface hydrological regime sufficiently to mix nutrients, solvents, and microbes. In addition, the costs of nutrients like those listed in Table 10.5, probably will not be significant, in most cases, although electron donors (e.g. methanol) and acceptors (e.g. hydrogen peroxide) might have significant costs associated with them.

As described previously, this natural bioremediation has the potential to reduce the concentrations of chlorinated solvents below detection levels, but would probably require either permission to reinject/recirculate contaminated water or to leave the site to slowly proceed without intervention, assuming adequate anaerobic and aerobic zones.

10.12. KNOWLEDGE GAPS

Unfortunately, without expanding our knowledge regarding the general ability of anaerobes to reductively dechlorinate those solvents, the effects of temperature, pH, ionic strength, soil or aquifer type on these subsurface microbial communities and the environmental distribution of the appropriate microbes, further development of this technology will be limited to sites where intensive treatability studies have shown it to be appropriate. Indigenous anaerobic microbes' dechlorination abilities and kinetic coefficients need to be evaluated. Implementation of this technology at study sites where conditions seem appropriate will aid in developing experience.

Clearly, research to date has demonstrated the potential for both anaerobic and aerobic degradation or transformation of chlorinated solvents in laboratory studies. However, research that addresses the problems associated with moving a technology from well-controlled environments into nature is lacking for these processes. Some of these studies must be undertaken by microbial ecologists who can evaluate both the biological potential and spatial distribution in nature. Other studies need to be performed by hydrogeologists who can evaluate the ability to induce anaerobic conditions throughout a contaminated site. Finally, risk associated with the natural bioremediation, both the process and possible products, needs to be evaluated.

10.13. CONCLUSION

When a chlorinated solvent is introduced into the environment, it may be transformed by chemical and biological reactions into a variety of products. These may be more or less hazardous than the original chlorinated solvent. Although chemical transformations may be quite slow, biological reactions often proceed quickly. The types of microbial conversions and the resulting products will depend on the chlorinated solvent and environmental conditions. An understanding of these transformations provides an insight into the natural processes and methods for producing conditions that will maximize degradation rates and lead to the conversion of chlorinated solvents to compounds that are not hazardous to human health. One such process might be anaerobic/aerobic degradation of chlorinated solvents.

REFERENCES

Anders, M.W. 1983. Bioactivation of halogenated hydrocarbons. *J. Toxicol.-Clin. Toxicol*. 19:699-706.

Assaf-Anid, N. L. Nies, and T.M. Vogel. 1992. Reductive dechlorination of a polychlorinated biphenyl congener and hexachlorobenzene by vitamin B12. *Appl. Environ. Microbiol*. 58(3):1057-1060.

Bagley, D.M., and J.M. Gossett. 1990. Tetrachloroethene transformation to trichloroethene and *cis*-1,2-dichloroethene by sulfate-reducing enrichment cultures. *Appl. Environ. Microbiol*. 56(8):2511-2516.

Bhatnagar, L., and B.Z. Fathepure. 1991. Mixed cultures in detoxification of hazardous wastes. In: *Mixed Cultures in Biotechnology*. Eds., G. Zeikus and E.A. Johnson. McGraw-Hill, Inc. New York. pp. 293-340.

Bouwer, E.J., and J.P. Wright. 1988. Transformations of trace halogenated aliphatics in anoxic biofilm columns. *J. Contamin. Hydrol*. 2:155-169.

Bouwer, E.J., and P.L. McCarty. 1985. Utilization rates of trace halogenated organic compounds in acetate-grown biofilms. *Biotech. Bioengr*. 27:1564-1571.

Brunner, W., D. Staub, and T. Leisinger. 1980. Bacterial degradation of dichloroethane. *Appl. Environ. Microbiol*. 40(5):950-958.

de Bont, J.A.M., M.J.A.W. Vorage, S. Hartmans, and W.J.J. van den Tweel. 1986. Microbial degradation of 1,3-dichlorobenzene. *Appl. Environ. Microbiol*. 52(4): 677-680.

Fathepure, B.Z., J.M. Tiedje, and S.A. Boyd. 1988. Reductive dechlorination of hexachlorobenzene to tri- and dichlorobenzenes in anaerobic sewage sludge. *Appl. Environ. Microbiol*. 53(2):330-347.

Fathepure, B.Z., and T.M. Vogel. 1991. Complete degradation of polychlorinated hydrocarbons by a two-stage biofilm reactor. *Appl. Environ. Microbiol*. 57(12):3418-3422.

Fogel, M.M., A.R. Taddeo, and S. Fogel. 1986. Biodegradation of chlorinated ethenes by a methane utilizing mixed culture. *Appl. Environ. Microbiol*. 51(4):720-724.

Freedman, D.L., and J.M. Gossett. 1989. Biological reductive dechlorination of tetrachloroethylene and trichloroethylene to ethylene under methanogenic conditions. *Appl. Environ. Microbiol*. 55(9):2144-2151.

Gantzer, C.J., and L.P. Wackett. 1991. Reductive dechlorination catalyzed by bacterial transition-metal coenzymes. *Environ. Sci. Technol*. 25(4):715-722.

Gibson, S.A., and Suflita, J.M. 1986. Extrapolation of biodegradation results to groundwater aquifers: Reductive dehalogenation of aromatic compounds. *Appl. Environ.Microbiol*. 52(4):681-688.

Henson, J.M., M.V. Yates, J.W. Cochran, and D.L. Shackleford. 1988. Microbial removal of halogenated methanes, ethanes, and ethylenes in aerobic soil exposed to methane. *FEMS Microbiol. Ecol.* 53:193-201.

Kuhn, E.P., P.C. Colberg, J.L. Schnoor, O. Wanner, A.J.B. Zehnder, and R.P. Schwarzenbach. 1985. Microbial transformations of substituted benzenes during infiltration of river water to groundwater: Laboratory column studies. *Environ. Sci. Technol.* 19(10):961-968.

Little, C.D., A.V. Palumbo, S.E. Herbes, M.E. Lidstrom, R.L. Tyndall, and P.J. Gilmer. 1988. Trichloroethylene biodegradation by a methane-oxidizing bacterium. *Appl. Environ. Microbiol.* 54(4):951-956.

Mabey, W.R., V. Barich, and T. Mill. 1983. Hydrolysis of polychlorinated alkanes, American Chemical Society, Div. Environ. Chem. Annual Meeting. September, Washington DC. pp. 359-361.

Mabey, W., and T. Mill. 1978. Critical review of hydrolysis of organic compounds in water under environmental conditions. *J. Phys. Chem. Ref. Data.* 7:383-415.

Mackay, D.M., and J.A. Cherry. 1989. Groundwater contamination: Pump-and-treat remediation. *Environ. Sci. Technol.* 23:630-636.

McCarty, P.L., and J.T. Wilson. 1992. Natural anaerobic treatment of a TCE plume St. Joseph, Michigan, NPL Site. In: *Abstracts, Symposium on Bioremediation of Hazardous Wastes. EPA's Biosystems Technology Development Program.* U.S. Environmental Protection Agency.

McCarty, P.L. 1984. Application of Biological Transformations in Ground Water. In: *Proceedings 2nd Int. Conf. on Ground Water Quality Research.* Tulsa, Oklahoma.

Merian, E., and M. Zander. 1982. Volatile aromatics. p. 117-161. In: *The Handbook of Environmental Chemistry.* Ed., O. Hutzinger. Vol. 3, Part B. Springer-Verlag. New York.

Nelson, M.J.K., S.O. Montgomery, E.J. O'Neille, and P.H. Pritchard. 1986. Aerobic metabolism of trichloroethylene by a bacterial isolate. *Appl. Environ. Microbiol.* 52(2):383-384.

Nelson, M.J.K., S.O. Montgomery, W.H. Mahaffey, and P.H. Pritchard. 1987. Biodegradation of trichloroethylene and involvement of an aromatic biodegradative pathway. *Appl. Environ. Microbiol.* 53(5):949-954.

Nelson, M.J.K., S.O. Montgomery, and P.H. Pritchard. 1988. Trichloroethyelene metabolism by microorganisms that degrade aromatic compounds. *Appl. Environ. Microbiol.* 54(2):604-606.

Nies, L., and T.M. Vogel. 1990. Effects of organic substrates on dechlorination of aroclor 1242 in anaerobic sediments. *Appl. Environ. Microbiol.* 56(9):2612-2617.

Pearson, C.R. 1982. Halogenated aromatics. In: *The Handbook of Environmental Chemistry.* Ed., O. Hutzinger. Vol. 3, Part B. Springer-Verlag. New York. pp. 89-116.

Phelps, T.J., J.J. Niedzielski, R.M. Schram, S.E. Herbes, and D.C. White. 1990. Biodegradation of trichloroethylene in continuous-recycle, expanded-bed bioreactors. *Appl. Environ. Microbiol.* 56(6):1702-1709.

Schraa, G., M.L. Boone, M.S.M. Jetten, A.R.W. van Neer ven, P.C. Colberg, and A.J.B. Zehnder. 1986. Degradation of 1,4-dichlorobenzene by *Alcaligenes* sp. strain A175. *Appl. Environ. Microbio*l. 52(6):1374-1381.

Spain, J.C., and S.F. Nishino. 1987. Degradation of 1,4-dichlorobenzene by a *Pseudomonas* sp. *Appl. Environ. Microbiol.* 53(5):1010-1019.

Strand, S.E., and L. Shippert. 1986. Oxidation of chloroform in an aerobic soil exposed to natural gas. *Appl. Environ. Microbiol.* 52(1):203-205.

Symons, J.M., T.A. Bellar, J.K. Carswell, J. Demarco, G.G. Kropp, D.R. Robeck, C.J. Seeger, B.L. Slocum, K.L. Smith, and A.A. Stevens. 1975. National organics reconnaissance survey for halogenated organics. *J. Amer. Water Works Assoc.* 67:634-647.

Tiedje, J.M., S.A. Boyd, and B.Z. Fathepure. 1987. Anaerobic biodegradation of chlorinated aromatic hydrocarbons. *Dev. Ind. Microbiol*. 27:117-127.

van der Meer, J.R., W. Roelofsen, G. Schraa, and A.J.B. Zehnder. 1987. Degradation of low concentrations of dichlorobenzenes and 1,2,4-trichlorobenzene by *Pseudomonas* sp. strain p51 in non-sterile soil columns. *FEMS Microbiol. Ecol.* 45:333-341.

Vogel, T.M. and P.L. McCarty. 1985. Biotransformation of tetrachloroethylene to trichloroethylene, dichloroethylene, vinyl chloride, and carbon dioxide under methanogenic transformation. *Appl. Environ. Microbiol.* 49(5):1080-1083.

Vogel, T.M., and M. Reinhard. 1986. Reaction products and rates of disappearance of simple bromoalkanes, 1,2-dibromopropane and 1,2-dibromoethane in water. *Environ. Sci. Technol.* 20(10):992-997.

Vogel, T.M., and P.L. McCarty. 1987a. Abiotic and biotic transformations of 1,1,1-tricholorethane under methanogenic conditions. *Environ. Sci. Technol.* 21(12):1208-1213.

Vogel, T.M., and P.L. McCarty. 1987b. Rate of abiotic formation of 1,1-dichloroethylene from 1,1,1-trichloroethane in groundwater. *J. Contam. Hydrol*. 1:299-308.

Vogel, T.M., C.S. Criddle, and P.L. McCarty. 1987. Transformations of halogenated aliphatic compounds. *Environ. Sci. Technol.* 21(8):722-736.

Vogel, T.M. 1988. *Biotic and Abiotic Transformations of Halogenated Aliphatic Compounds*. Ph.D. Thesis. Stanford University. Stanford, California.

Walton, B.T., and T.A. Anderson. 1990. Microbial degradation of trichloroethylene in the Rhizosphere: Potential application to biological remediation of waste sites. *Appl. Environ. Microbiol.* 56(4):1012-1016.

Weintraub, R.A., G.W. Jex, and H.A. Moye. 1986. Chemical and microbial degradation of 1,2-dibromoethane (EDB) in Florida ground water, soil, and sludge. In: *Evaluation of Pesticides in Ground Water*. Eds., W.Y. Garner, R.C. Honeycutt, and H.N. Nigg. American Chemical Society. Washington, DC. pp. 294-310.

Westrick, J.J., W. Mello, and R.F. Thomas. 1984. The groundwater supply survey. *J. Amer. Water Works Assoc.* 76(5):52-59.

Zeikus, J.G., J.A. Kerby, and J.A. Krzycki. 1985. Single-carbon chemistry of acetogenic and methanogenic bacteria. *Science*. 227:1167-1173.

OTHER REFERENCES

Bishop, P.L., and N.E. Kinner. 1986. Aerobic fixed-film process. In: *Biotechnology. Vol. 8.* Eds., H.J. Rehm and G. Reed. VCH Verlagsgesell-schaft mbH, D-6940, Weinheim, Fed. Rep. Germany.

Cunningham, A.B., E.J. Bouwer, and W.G. Characklis. 1990. Biofilms in porous media. In: *Biofilms.* Eds., W.G. Characklis and K.C. Marshall. John Wiley and Sons, Inc. New York. pp. 697-732.

Fathepure, B.Z. 1987. Factors affecting the methanogenic activity of *Methanothrix soehngenii* VNBF. *Appl. Environ. Microbiol.* 53(12):2978-2982.

Rittmann, B.E. 1987. Aerobic biological treatment. *Environ. Sci. Technol.* 21(2):128-136.

Criddle, C.S., P.L. McCarty, M.C. Elliott, and J.F. Barker. 1986. Reduction of hexachloroethane to tetrachloroethylene in groundwater. *J. Contam. Hydrol.* 1:133-142.

Hallen, R.T., J.W. Pyne, Jr., and P.M. Molton. 1986. Transformation of chlorinated ethenes and ethanes by anaerobic microorganisms. *Amer. Chem. Soc. Ann. Mtg., Extended Abstract.* 344-346.

SECTION 11

INTRODUCED ORGANISMS FOR SUBSURFACE BIOREMEDIATION

J. M. Thomas and C. H. Ward
Rice University
National Center for Ground Water Research
Department of Environmental Science and Engineering
Houston, Texas 77251
Telephone: (713)527-4086
Fax: (713)285-5203

11.1. FUNDAMENTAL PRINCIPLES OF THE TECHNOLOGY

Inocula of microorganisms have been widely used for bioremediation of hazardous waste sites. There is little documentation of the efficacy of this process, and important questions still persist about the environmental responsibility of adding nonindigenous microorganisms. This section reviews the properties of the subsurface and the properties of microorganisms that influence their transport through geological material, their survival, and their capacity to degrade contaminants.

11.1.1. Review of the Development of the Technology

The concept of microbial movement through the subsurface was first addressed as early as the mid-1920s for microbial enhanced oil recovery (MEOR). At that time, Beckmann (1926) suggested that microorganisms that produce emulsifiers or surfactants could be transported into an oil-bearing formation to recover oil that remains after a well has stopped flowing. The addition of microorganisms to oil-bearing formations to enhance oil recovery by biosurfactant or biogas production has since been investigated and appears promising (Bubela, 1978). At about the same time, research on the transport of microorganisms through the subsurface environment was being conducted to determine the effectiveness of on-site wastewater disposal systems (i.e. pit latrines, septic tanks, land disposal of sewage) in removing pathogens (Caldwell, 1937, 1938). More recently, the concept of transporting microorganisms with specialized metabolic capabilities for subsurface bioremediation has been proposed (Lee et al., 1988; Thomas and Ward, 1989).

The addition of microorganisms to the subsurface in remedial operations would be beneficial when contaminants resist biodegradation by the indigenous microflora, where evidence of toxicity exists, or when the subsurface has been sterilized by the contamination event. Seed microorganisms have been added to the subsurface to aid in contaminant biodegradation; however, the role of the added microorganisms has never been differentiated from that of the indigenous microflora (Lee et al., 1988; Thomas and Ward, 1989). Operations in which seed organisms are added to enhance contaminant biodegradation in the subsurface usually involve treating contaminated ground water in a closed-loop system by withdrawal and treatment in an aboveground bioreactor or by physical methods, after which the treated ground water is reinjected into the subsurface. The treated ground water that is reinjected contains adapted microorganisms from the bioreactor or is amended with contaminant-degrading organisms to enhance biodegradation in situ (Ohneck and Gardner, 1982; Quince and Gardner, 1982a, b; Winegardner and Quince, 1984; Flathman and Githens, 1985; Flathman and Caplan, 1985, 1986; Flathman et al., 1985; Quince et al., 1985).

For added microorganisms to be effective in contaminant degradation, they must be transported to the zone of contamination, attach to the subsurface matrix, survive, grow, and maintain their degradative capabilities (Thomas and Ward, 1989). When injected into a nonsterile formation, the added organisms must compete with the indigenous microflora for limiting nutrients and escape predation. Transport will depend upon complex interactions between the subsurface and the microorganism. Physical phenomena related to the composition of the subsurface formation that affect transport include filtration and adsorption (Gerba and Bitton, 1984). Passage through the subsurface will depend on grain size and related values of hydraulic conductivity (K) and channels made by cracks and fissures. However, transport by channeling probably will result in uneven seeding of the formation. Obviously, organisms that are larger than the average pore size cannot move with ground water and will be retained by aquifer solids. In addition, microbial cells may be removed from solution by sorption in sediments high in clay and organic matter.

One of the first studies that addressed microbial transport through subsurface materials for the purpose of contaminant degradation was published by Raymond et al. (1977). These investigators reported that heterotrophs and hydrocarbon-degrading bacteria penetrated and were detected in the effluent of 1.45 x 31 cm columns packed with unconsolidated sands having effective hydraulic conductivity (K) values ranging from 3.38×10^{-3} to 1.9×10^{-1} cm/sec, which were run at a flow rate of about 30 ml/h (Darcy flow 18/cm/hr). Microorganisms also penetrated and were detected in the effluent of 3.8 x 10 cm sandstone (consolidated) cores, with hydraulic conductivities ranging from 1.8×10^{-5} to 7.2×10^{-5} cm/sec, through which water was passed under unknown pressure. In a separate experiment, it was determined that the added microorganisms were utilizing the gasoline.

11.1.2. Matrix Properties That Affect Transport

Bioremediation of the subsurface is usually limited in subsurface material with hydraulic conductivities less than 10^{-4} cm/sec, because of the difficulty in pumping fluids through material with lower K values. Hence, for practical purposes, microbial transport through the subsurface to enhance bioremediation will probably be limited to material with hydraulic conductivities of 10^{-4} cm/sec or greater. Laboratory studies using materials that have been screened or sieved has produced a distorted view of microbial transport in geological materials; the transport of microbial cells with the flow of ground water has been underestimated. In-situ geological materials have a more heterogeneous distribution of pore sizes than laboratory simulations. As a practical consequence, microbial inocula can move readily through the larger pores in many subsurface environments.

Several studies have been conducted to determine the effect of hydraulic conductivity on microbial transport for selective plugging of subsurface materials for MEOR. Hart et al. (1960) reported that the plugging effect of injecting water containing 1.2×10^6 dead bacteria/ml through 2.5 x 8 cm sandstone (consolidated) cores maintained at an input pressure of 40 psi were not different at hydraulic conductivities ranging from 1.2 to 2.9×10^{-4} cm/sec. Kalish et al. (1964) found that when 1×10^6 dead cells/ml of a *Pseudomonas aeruginosa* strain was transported through 2.54 x 5.08 to 10.16 cm sandstone cores at constant flow rate, core plugging was inversely related to K at hydraulic conductivities ranging from 3×10^{-5} to 2.8×10^{-4} cm/sec. Bubela (1978) reviewed several papers on MEOR and found that secondary recovery in formations with K values of 9.66 x 10^{-5} cm/sec or greater could be enhanced using microbiological techniques; however, recovery declined or was insignificant in less permeable formations. Jenneman et al. (1985) found that the rate of penetration of a motile *Bacillus* sp. through nutrient-saturated sandstone cores under static conditions was independent of hydraulic conductivity at values above 9.66 x 10^{-6} cm/sec but rapidly decreased for cores with K values below it.

More recent studies have been conducted to determine the effect of K on microbial transport for enhancement of subsurface bioremediation. Marlow et al. (1991) reported that extent of transport of a yeast, *Rhodotorula* sp., after 10 pore volumes through sand columns with hydraulic conductivity values of 5.59 x 10^{-2} and 1.37 x 10^{-1} cm/sec was about 2 and 50%, respectively, of the initial number of cells added (1 to 2 x 10^5 cells/ml). Fontes et al. (1991) investigated the effects of grain size, bacterial cell size, ionic strength of the transporting fluid, and heterogeneities of the medium on microbial transport and found that grain size was the most important variable. When transporting a gram-negative coccus in a low ionic strength fluid at a flow rate of 88 ml/h (Darcy flow 4.9 cm/hr) through sand packed in a 4.8 x 14 cm column to achieve K values of 2.0 cm/sec and 0.37 cm/sec, 88.35 and 14.50% of the cells, respectively, were recovered in the effluent.

The mineralogy of the matrix may also affect transport. Scholl et al. (1990) found that bacterial attachment, and hence removal from solution, may be affected by the surface charge on the minerals present. These authors reported that attachment of bacteria (negatively charged) isolated from ground water was greater for limestone, iron hydroxide coated quartz and iron hydroxide coated muscovite (positively charged) than for clean quartz and clean muscovite (negatively charged) in batch experiments. Attachment to coated muscovite was greater than that to coated quartz whereas attachment to clean muscovite and quartz was not different. When bacterial cells (1.77 x 10^9 cells/ml) were transported through columns (2 x 20 cm) packed with coated or uncoated quartz at a flow rate of 21 ml/h (Darcy flow 6.7 cm/hr), 99.9 and 97.4% of the cells, respectively, were retained in the columns.

Another matrix property that affects transport is sediment structure. Smith et al. (1983) reported that *Escherichia coli* was transported to a greater extent through intact cores than through cores of disturbed or structureless soil. In addition, movement through intact cores appeared to be related to the presence of macropores and channeling. For intact cores, there was no relationship between clay and organic matter content and the extent of transport. These authors suggested that the use of studies in which the porous material is sieved and then packed homogeneously in columns to determine microbial transport will not be predictive of transport in situ because the natural pores and channels will be destroyed.

Madsen and Alexander (1982) reported that vertical transport of *Rhizobium japonicum* and *Ps. putida* was facilitated by percolating water, plant roots and percolating water, and a burrowing earthworm. Neither species of bacteria was transported further than 2.7 cm below the surface without facilitation. Transport through channels was thought to be the most important mechanism for microbial movement.

To summarize, matrix properties that will affect transport include hydraulic conductivity, mineralogy, and sediment structure. Hydraulic conductivity was the most studied parameter affecting transport through porous media; however, the results of laboratory studies in which samples of porous media were packed to homogeneity may produce underestimates of microbial transport. The use of intact cores will provide the full range of pore sizes present in situ for microbial transport through available macropores.

11.1.3. Properties of Organisms that Affect Transport

Properties of organisms that affect transport include size, shape, stickiness, condition, and motility and chemotaxis. In general, no one property has a dominant influence on transport of microbial cells. The relative influence of organismal properties is less important than the properties of the geological matrix, or operational factors such as cell density, or chemical properties of the ground water.

Kalish et al. (1964) investigated the effect of cell size and aggregation tendencies of *Ps. aeruginosa* (0.5 x 1.5 μm; no aggregation), *Micrococcus roseus* (0.8 μm; occurs singly or as aggregates), and *B. cereus* (1 x 6 μm; occurs singly or as long chains) on their ability to plug sandstone cores with high (2.6 to 3.3 x 10^{-4} cm/sec) and low (2.1 to 2.9 x 10^{-5} cm/sec) hydraulic conductivity at initial cell densities of 1 x 10^5 and 1 x 10^4 dead cells/ml, respectively. The authors found that the aggregation tendency of the cells was more important than cell size in causing a reduction in hydraulic conductivity. *Ps. aeruginosa*, which is intermediate in size, caused the least amount of plugging; *M. roseus* which is the smallest and occurs as single cocci or as aggregates, caused more plugging, while *B. cereus*, which is the largest and occurs as single rods or in chains, caused the most plugging. However, when *Ps. aeruginosa* and another nonaggregating but larger bacterium, *Proteus vulgaris* (0.5 to 1.0 x 1 to 3 μm) were transported through sandstone cores with similar hydraulic conductivity, the larger organism, *P. vulgaris*, caused the most plugging.

Gannon et al. (1991) investigated the transport of the rod-shaped organisms *Enterobacter*, *Pseudomonas*, *Bacillus*, *Achromobacter*, *Flavobacterium*, and *Arthrobacter* strains through 10 x 5 cm columns packed with a loam soil at a flow rate of 2.5 cm/h; cells with lengths less than 1 μm were transported to a greater extent than larger cells. The presence of capsules and the hydrophobic nature and the net surface electrostatic charge of the cell, properties that may affect sorption, did not influence transport.

Fontes et al. (1991) transported a gram-negative coccus (approximately 0.75 μm in diameter) and gram-negative rod (approximately 0.75 x 1.8 μm) with similar hydrophobicities through columns packed with unconsolidated sand with hydraulic conductivities of 0.37 and 2.0 cm/sec at a flow rate of 88 ml/h; the coccus was transported to a greater extent than the rod. Jang et al. (1983) transported *Ps. putida*, *Clostridium acetobutylicum* spores, and vegetative cells and spores of *B. subtilis* through 2.54 x 7.62 cm sandstone cores with a hydraulic conductivity of 3.9 x 10^{-3} cm/sec at a flow rate of 40 ml/h (Darcy flow 7.9 cm/hr); *Cl. acetobutylicum* spores were transported to the greatest extent. In addition, *B. subtilis* spores were transported to a greater extent than were the vegetative cells. Another property that may affect transport is the condition of the cell. MacLeod et al. (1988) reported that starved cultures of *Klebsiella pneumoniae*, which were smaller and less sticky than vegetative cells, were transported through artificial (glass beads) rock cores to a greater extent than the vegetative cells.

Motility and chemotaxis (the ability of a cell to detect and move with substrate gradients) may be important in the movement of microorganisms to contaminants localized in the subsurface. Jenneman et al. (1985) compared two taxonomically similar strains, *En. aerogenes*, which is motile, and *K. pneumonia*, which is nonmotile. The motile strain penetrated nutrient-saturated sandstone cores of similar length and hydraulic conductivities (4.5 to 6.1 x 10^{-4} cm/sec) 3 to 8 times faster than the nonmotile strain under nonflow conditions. In addition, penetration of either strain was not related to hydraulic conductivity within the ranges tested.

Using isogenic strains of *E. coli* under anaerobic conditions, researchers from this same laboratory found that motility, but not chemotaxis, may be important in the penetration of cells through unconsolidated porous media (Reynolds et al., 1989). Mutant nonchemotactic motile strains penetrated 2.01 x 8 cm nutrient-saturated sand cores faster than a chemotactic motile strain (wildtype) and nonmotile mutants, under static conditions. In addition, a motile strain that was chemotactic toward but unable to utilize galactose, penetrated sand cores at the same rate in the presence or absence of a galactose gradient. Furthermore, nonmotile strains of *E. coli* which produced gas penetrated nutrient-saturated cores about 5 to 6 times faster than mutant nongas-producing strains, suggesting that gas production is important for movement of nonmotile cells. For both motile and nonmotile strains, penetration rates were directly related to growth.

To summarize, the properties of the microorganisms that may affect transport include size, aggregating tendencies, shape, condition, and motility and chemotaxis. The results of studies designed to investigate which cell characteristics are most important are mixed. Cell size is important in that transport will be limited or prevented for cells that are bigger than the average pore size; however, cells that tend to aggregate, even if they are small, will not be good candidates for transport. For microorganisms that form spores, the spore, which is smaller than the vegetative stage, may be transported more efficiently. Microorganisms that are in a starved state usually are smaller and produce less extracellular polysaccharide, which allows the organism to attach to surfaces. Thus the reduced size and stickiness of the cells should enhance transport. Finally, motility may enhance transport.

11.1.4. Operational Factors that Affect Transport

The most important operational factor is the ionic strength of the water used to introduce the microorganisms into the subsurface. Transport is greatly facilitated in water with low ionic strength. The concentration of microbial cells or spores may also affect the rate and extent of transport through subsurface materials. At high cell densities, filtration of cells can significantly reduce hydraulic conductivity. At low cell density, the cells sorb to aquifer matrix materials to a greater extent. Transport is also related to the rate of flow of water. A greater proportion of cells are transported in water that is moving rapidly.

Hart et al. (1960) reported on the effect of injection concentration of cells on hydraulic conductivity of consolidated sandstone cores (2.5 x 8 cm; $K = 9.66 \times 10^{-5}$ cm/sec) maintained at an input pressure of 40 psi. Hydraulic conductivity decreased as the injection concentration of cells increased from 1.2×10^5 to 1.2×10^7 cells/ml. Kalish et al. (1964) also found that the injection concentration of dead cells of *Ps. aeruginosa* and *M. luteus* at concentrations ranging from 1×10^6 to 20×10^6 and 1×10^5 to 1×10^6 cells/ml, respectively, was inversely related to the final hydraulic conductivity of sandstone cores through which the cells were transported at constant flow rate (initial $K = 3 \times 10^{-4}$ cm/sec). The same trend of decreasing K with increasing cell concentration was observed when *Ps. aeruginosa* was transported through high (3×10^{-4} cm/sec) and low (1.5×10^{-5} cm/sec) permeability sandstone. At influent concentrations of 5×10^7 but not 1×10^6 cells/ml, Jang et al. (1983) observed formation of a filter cake at the inlet and a pressure drop along the core when *Ps. putida* was injected under nongrowth conditions through 2.54 x 7.62 cm sandstone cores ($K = 3.9 \times 10^{-3}$ cm/sec; flow rate of 40 ml/h). A filter cake did not form at influent concentrations of 1×10^6 cells/ml.

In contrast to studies that indicate an inverse relationship between transport and cell density, Smith et al. (1983) found that cell densities ranging from 10^5 to 10^8 cells/ml had no effect on the extent of transport of *E. coli* through several different intact cores. These authors speculated that these relatively large concentrations of cells could not saturate the adsorption and filtration sites of these soils. Reynolds et al. (1989) found that the penetration rate of an *E. coli* strain through 2.01 x 8 cm sand cores increased in proportion to the logarithm of cell concentration. Bacterial concentrations between 10^1 to 10^7 cells/ml were tested. These authors speculated that factors that may inhibit cell movement through porous media on a finite basis, such as sorption, may be negated by increasing the initial number of cells added.

Gannon et al (1991) also found a direct relationship between transport and cell density. Bacterial cells were transported in deionized water or a 0.01 M NaCl solution through columns (5 x 30 cm) packed with sandy aquifer material at a Darcy flow rate of 10^{-4} cm/sec. An increase in injected cell density from 10^8 to 10^9 cells/ml increased the total recovery of cells transported in deionized water and the salt solution from 44 to 57% and 1.5 to 44%, respectively. The authors suggested that the small increase in recovery of cells injected in deionized water resulted from the lack of appreciable adsorption under conditions of low ionic strength. However, the large difference in recovery of cells transported in the salt solution, a condition of high ionic strength which favors sorption, was the result of a smaller percentage of cells at the higher density that was retained by a finite number of sorption sites.

Although injection concentration may physically retard or enhance transport, injection concentration or inoculum size will be important in initiation and maintenance of biodegradation. Ramadan et al. (1990) found that inoculum size affected the biodegradation potential of bacteria inoculated into lake water. Inoculation of *Ps. cepacia* into lake water resulted in mineralization of 1 µg/ml *p*-nitrophenol at concentrations of 3.3 x 10^4 and 3.6 x 10^5 cells/ml but not at 330 cells/ml. The absence of biodegradation at low cell concentrations was a result of protozoa that grazed the population to nondetectable levels. Grazing by protozoa could significantly reduce the number of naturally occurring and/or introduced contaminant-degrading organisms and affect the rate and extent of bioremediation. Sinclair (1991) reported that 100 or less eucaryotes were detected in samples from two uncontaminated sites; however, large numbers of protozoa (2.66 x 10^5/gram dry weight) were detected in samples contaminated with jet fuel, aviation fuel, and creosote in which sufficient organic carbon was present to support high numbers of bacteria (Sinclair, 1991; Madsen et al., 1991).

Flow rate is another factor that may affect transport of microorganisms through the subsurface. In all published reports, an increase in flow rate increased transport. Kalish et al. (1964) found that reductions in hydraulic conductivity in sandstone as a result of plugging by suspensions of dead cells of *P. vulgaris* could be partially reversed by increasing the flow rate, which concomitantly increased the pressure differential across the core. Smith et al. (1983) reported a direct relationship between flow rate and the extent of transport of *E. coli* through intact cores. By increasing the flow rate from 0.5 to 4 cm/h, the extent of transport in a silt loam increased six times. Marlow et al. (1991) reported that transport of *Rhodococcus* sp. through sand packs with K = 1.37 x 10^{-1} cm/sec was facilitated by increasing the injection rate; increasing the flow rate by a factor of two nearly doubled the number of cells transported through the column. Gannon et al. (1991) transported bacterial cells (10^8 cells/ml) in deionized water or a 0.01 M NaCl solution through a column (5 x 30 cm) packed with sandy aquifer material. Flow rates ranged from 1 x 10^{-2} to 2 x 10^{-2} cm/sec. Doubling the rate of flow increased the total recovery of cells transported in deionized water and on 0.01 M NaCl solution, respectively, from 60 to 77% and from 1.5 to 3.9%.

In summary, the operational factors that will affect microbial transport include cell concentration, flow rate, and the ionic strength of the transporting fluid. The results of studies designed to investigate the effects of cell density on transport have been mixed. The effects of cell density on transport may be organism- and site-specific. Microbial filtration and clogging of the matrix will be of concern. The direct relationship between flow rate and microbial transport always has been found. Finally, there will be an inverse relationship between the ionic strength of the transporting fluid and microbial transport. Microorganisms tend to sorb to surfaces under conditions of high ionic strength and to a less extent under condition of low ionic strength; hence more cells will be transported in a fluid of low ionic strength.

11.1.5. Environmental Factors that Affect Survivability of Added Organisms

As mentioned previously, transported organisms must not only reach the zone of contamination but must compete with the indigenous microflora for nutrients, escape predation, retain their biodegradative capabilities, and often tolerate extremes in pH, temperature, and other environmental variables. Hardly anything is known about environmental factors and survivability in the subsurface environment. By extrapolation from experience with surface water and soil, predation will probably be the most important factor limiting the survival and activity of introduced microorganisms.

Goldstein et al. (1985) reported that the success of adding nonindigenous microorganisms to the environment may be dependent on the concentration of the target compound, the presence of toxicants or predators, the preferential use of alternate substrates, or the mobility of the introduced organisms. In one experiment, these authors found that mixing enhanced the mineralization of 5 µg/g *p*-nitrophenol (PNP) in sterile soil inoculated with a PNP-degrading organism, suggesting that mixing was required to move the organisms through the soil to effectively degrade the PNP. Zaidi et al. (1988) reported that pH and substrate concentration affected the survival and biodegradation capabilities of introduced organisms in lake water. These authors found that an increase in pH from 7 to 8 inhibited the mineralization of PNP in sterile and nonsterile lake water inoculated with a *Pseudomonas* sp.

The presence of predators and inhibitors may also affect the survival and biodegradation potential of inoculants. Zaidi et al. (1989) found that the addition of a eucaroytic inhibitor to lake water inoculated with a *Corynebacterium* sp. increased the extent of mineralization of 26 ng/ml PNP, but did not increase mineralization of higher concentrations of PNP; the authors suggested that the organisms were not able to replace those cells cropped by eucaryotic grazing at the lower concentrations of PNP.

In summary, the same factors that affect survival of microorganisms in the surface soil and water environments will affect the survivability in the subsurface. These factors include substrate concentrations, pH, temperature, and the presence of toxicants, predators, and alternate substrates. However, little information is available concerning the survivability of introduced microorganisms in the subsurface.

11.1.6. Field Demonstrations of Microbial Transport

Field demonstrations have documented the transport of introduced microorganisms through the subsurface. In one demonstration, bacteria native to the aquifer actually moved faster than the bulk flow of ground water, perhaps due to size exclusion chromatography.

In 1978, Hagedorn et al. (1978) transported antibiotic-resistant fecal bacteria through subsoil under saturated conditions at depths of 30 to 60 cm below the surface to investigate the potential for ground-water contamination by septic tank discharge. When inocula at concentrations ranging from 3 to 5 x 10^8 cells/ml were added, bacteria were detected in sampling wells located 50 cm from the injection point after 1 day. In some wells, cells were detected as far as 1,500 cm from injection after 8 and 12 days. Microbial numbers peaked in sampling wells after rainfall, suggesting that transport was associated with rainfall patterns. Researchers from this same laboratory investigated the transport of antibiotic-resistant *E. coli* through different horizons of hillslope soils under saturated conditions (Rahe et al., 1978a, 1978b). Inocula at a concentration of 1.4 x 10^9 cells/ml were injected into the subsoil at depths ranging from 12 to 80 cm, and their numbers monitored downslope at distances of 2.5 to 20 m from the injection point at depths ranging from 12 to 200 cm. Irrespective of inoculation depth, the cells moved downslope to zones of high permeability and then through macropores. In addition, transport was faster in a subsoil with greater slope and hydraulic conductivity than that with a lesser gradient and hydraulic conductivity.

In a similar study, McCoy and Hagedorn (1980) investigated transport under saturated conditions of antibiotic-resistant strains of *E. coli* through a concave hillslope, which was located in a transition area between two soil series. Inocula at 4 x 10^7 cells/ml were injected into horizontal injection lines located at depths of 12, 35, and 70 cm, and the numbers of transported bacteria were monitored at 2.5, 5.0, 10.0, and 15.0 m downslope at depths ranging from 12 to 200 cm. Bacterial transport varied with depth of injection in the upper soil series but not in the transition zone where flow paths converged. Flow in this zone resulted more from channeling rather than matrix flow as the water moved upward into more transmissive layers because of the hydraulic gradient and a nontransmissive clay layer.

Harvey et al. (1989) investigated the transport of bacteria and microspheres through a sandy aquifer (K = 0.1 cm/sec) in natural and forced gradient tracer experiments. The bacteria to be transported were cultured from ground water collected at the site and stained with a DNA-specific fluorochrome. A conservative tracer (Cl^- or Br^-), microspheres of different diameters (0.2 to 1.3 μm) and surface charges, and the indigenous bacteria (0.2 to 1.6 μm in length) were then injected into a well screened at 10 to 11 m below the surface and their transport was monitored at multilevel wells placed 1.7 and 3.2 m downgradient of the injection point.

In the forced gradient experiment, both bacteria and carboxylated microspheres were injected. Breakthrough of bacteria occurred somewhat earlier than that of bromide. The microspheres were retained by aquifer sediments to a greater extent than bacteria. Transport of microspheres was directly related to size.

In the natural gradient experiment, carboxylated microspheres of diameters ranging from 0.2 to 1.3 μm, uncharged microspheres with a diameter of 0.6 μm, and microspheres with a diameter of 0.8 μm and containing carbonyl surface groups were injected. Transport of the carboxylated microspheres was directly related to size. For the microspheres with different surface characteristics, increasing breakthrough times were observed for uncharged, carbonyl containing and carboxylated particles, respectively.

The results of these field demonstrations suggest that microorganisms can be transported significantly through subsurface materials. These data are contradictory to many laboratory experiments in which subsurface material was packed to achieve homogeneity and eliminate any macropores and channels that could have facilitated transport.

11.1.7. Inoculation to Enhance Biodegradation of Hydrocarbons

Microorganisms have been added to samples of soil and water in the laboratory and field to enhance biodegradation of hydrocarbons; however, the results of these studies have been mixed. Atlas (1977) stated in a review on stimulated petroleum biodegradation that seeding will not be necessary in most environments because of the ubiquity of hydrocarbon-degrading organisms. Although hydrocarbon-degrading organisms may be ubiquitous, the problem with natural bioremediation of these compounds is that the rate of biodegradation is often too slow. Nutrient addition and agents that render the compounds more bioavailable may enhance these rates. However, inoculation may be important in environments in which the population of hydrocarbon-degrading organisms is too low or absent, or the environment is too harsh. In the latter case, the added organisms must be able to tolerate the extreme conditions. In addition, inoculation may be beneficial in the biodegradation of the high-molecular-weight polycyclic aromatic hydrocarbons, which are recalcitrant (Bossert and Bartha, 1986). If seeding is considered as a method for hydrocarbon remediation, a mixture of microorganisms will be required. Zajic and Daugulis (1975) found that multiple species were required to degrade the complex composition of crude oil.

Several investigators have studied the effects of inoculants on hydrocarbon degradation. Schwendinger (1968) investigated the effect of adding nitrogen, phosphorus and a bacterial seed (*Cellulomonas* sp.) to reclaim soil contaminated with oil. Seeding did not enhance bioreclamation in soil amended with 25 ml/kg oil and inorganic nutrients; however, seeding did enhance bioreclamation in soil amended with 62 and 100 ml/kg oil and inorganic nutrients. Similarly, Jobson et al. (1974) also reported that a mixed population of hydrocarbon-degrading organisms slightly stimulated the degradation of the n-alkanes with chain lengths of C_{20} to C_{25} but had no effect on other components in soil amended with crude oil. Lehtomaki and Niemela (1975) reported that the addition of a mixture of hydrocarbon-degrading microorganisms to soil amended with 0.5% light fuel oil or heavy waste oil had no effect on oil decomposition. Westlake et al. (1978) added hydrocarbon-degrading bacteria to field plots amended with oil in the boreal region of the Northwest Territories and found that seeding did not enhance biodegradation above those plots which received fertilizer.

Several investigators have isolated organisms that can degrade the recalcitrant high-molecular-weight polycyclic aromatic hydrocarbons. Mueller et al. (1990) isolated a strain of *Ps. paucimobilis* from a creosote waste site that can metabolize several PAHs when its enzymes are induced by growth on fluoranthrene. The organism uses fluoranthrene, 2,3-dimethylnaphthalene, and phenanthrene, and to a lesser extent anthracene, benzo[b]fluorene, naphthalene, 1-methylnaphthalene, and 2-methylnaphthalene, as sole sources of carbon and energy. Washed cells of a fluoranthrene-grown culture were active against these compounds and biphenyl, anthraquinone, pyrene, and chrysene as well. The authors speculated that this organism may be effective in treating mixtures of PAHs, which are characteristic of creosote waste sites.

Heitkamp and Cerniglia (1988) isolated a *Mycobacterium* sp. from sediments exposed to petroleum hydrocarbons which was able to mineralize naphthalene, phenanthrene, fluoranthrene, pyrene, 1-nitropyrene, 3-methylcholanthrene, and 6-nitrochrysene when grown in the presence of peptone, yeast extract, and starch. Inoculation of soil and water mixtures with fluoranthrene-induced cells enhanced the mineralization of fluoranthrene by 93% over uninoculated samples (Kelley and Cerniglia, 1990). In addition, peptone, yeast extract, and starch amendments greatly enhanced the fluoranthrene-mineralization capability of the inoculant.

Microorganisms subjected to genetic manipulation may loose traits that allow them to survive and express the contaminant-degrading genes under conditions that prevail in the subsurface environment. A unique inoculation experiment involved the introduction of genetically modified ground-water bacteria, which harbored plasmids for toluene/xylene metabolism (TOL) and antibiotic resistance (RK2), into microcosms containing uncontaminated and artificially contaminated aquifer material (Jain et al., 1987). Although the inoculant was stably maintained for 8 weeks of incubation; inoculation did not enhance the biodegradation of toluene and chlorobenzene in the artificially contaminated microcosms above that of uninoculated microcosms.

In summary, inoculation to enhance biodegradation of most hydrocarbons usually is not necessary because of the ubiquity of hydrocarbon-degrading microorganisms. Microorganisms have coevolved with hydrocarbons, and many have metabolic capability to degrade the compounds. However, inoculation may be beneficial in environments in which there are harsh conditions or high molecular-weight polycyclic aromatic compounds.

11.1.8. Inoculation to Enhance Biodegradation of Chlorinated Compounds

In contrast to the situation with naturally occurring organics, inoculation to enhance the biodegradation of chlorinated compounds may be beneficial. This is particularly true for pentachlorophenol.

As early as 1965, MacRae and Alexander (1964) inoculated alfalfa seeds with a strain of *Flavobacterium* sp. that degraded 4-(2,4-dichlorophenoxy)- butyric acid, in order to protect the developing plants in herbicide-treated soils. Seeds were planted in nonsterile and sterile soils. Inoculation afforded protection in sterile soil but not in the presence of the indigenous microflora. Edgehill and Finn (1983) reported that the addition of 10^6 cells per g dry soil of a pentachlorophenol (PCP)-degrading strain of *Arthrobacter* reduced the half-life of 20 µg PCP/g soil from 2 weeks to 1 day in laboratory experiments; the bacterium used PCP as the sole source of carbon and energy. In addition, PCP biodegradation was directly related to inoculum size; PCP was reduced by 90% after 24, 40 and 100 h, after the addition of 10^6, 10^5 and 10^4 cells/g soil, respectively. The results of a field experiment conducted using soil in an outdoor shed indicated that inoculation and mixing enhanced PCP degradation. After 12 days, 25% was removed in uninoculated plots, 50% was removed in inoculated but unmixed plots, and 85% was removed in inoculated and mixed plots.

Martinson et al. (1984) inoculated samples of river water with 10^6 cells/ml of a PCP-degrading strain of *Flavobacterium* and found that 90% of the PCP (1 ppm) was removed within 48 h, whereas none was removed in uninoculated samples.

Investigators from this same laboratory investigated mineralization of PCP from contaminated soils by inoculation with the same PCP-degrading *Flavobacterium* (Crawford and Mohn, 1985). When samples of loam, clay and sand were amended with 100 ppm PCP and inoculated, initial rates of mineralization were initially fastest in the loam and slowest in the sand. However, about 60% of the PCP was mineralized in all soils after 6 days of incubation. PCP was not mineralized in uninoculated samples after 10 days. In another experiment in which samples were incubated for a longer period of time, significant mineralization of 100 ppm PCP was detected in uninoculated as well as inoculated samples, indicating that the indigenous microflora had acclimated to degrade PCP. After 40 days of incubation, PCP had been mineralized to the same extent (50%) in inoculated and uninoculated samples.

In one waste-dump soil contaminated with 298 ppm PCP, however, mineralization was detected in the inoculated samples only. In another waste-dump soil contaminated with 321 ppm PCP, the extent of removal was similar in inoculated and uninoculated samples. These authors speculated that enhancing PCP degradation by the indigenous microflora may be advantageous at low contaminant concentration while inoculation may be beneficial at high contaminant concentrations.

Brunner et al. (1985) investigated the effect of inoculation to enhance degradation of polychlorinated biphenyls in samples of soil. Soil (100 g) was adjusted to 50% water holding capacity, amended with 100 mg Aroclor 1242/kg, and inoculated with 10^5 or 10^9 cells/ml of a PCB-degrading *Acinetobacter*. After incubation aerobically for 70 days, inoculation did not greatly enhance mineralization; however, inoculation coupled with analog enrichment with biphenyl significantly enhanced mineralization above that observed in samples receiving biphenyl only.

Stormo and Crawford (1992) have developed a unique method for transporting a chlorophenol-degrading *Flavobacterium* sp. by encapsulating the cells into polymeric beads of alginate, agarose, or polyurethane. The cells are encapsulated into beads with diameters ranging from 2 to 50 μm. The catabolic activity of free and encapsulated cells is not different. Mineralization of PCP by free or encapsulated cells in 20 cm-columns containing native aquifer material has been assessed (Keith E. Stormo, personal communication). Preliminary results indicate that rates of PCP mineralization at concentrations as high as 200 mg/kg by free or encapsulated cells are not significantly different; however, encapsulation may enhance long-term survivability. PCP mineralization by the indigenous aquifer microflora was not observed.

To summarize, inoculation to enhance the biodegradation of chlorinated compounds may be beneficial in some instances. In contrast to the coevolution of microorganisms and hydrocarbons, the coexistence of microorganisms and chlorinated compounds has been relatively short. Many microorganisms cannot degrade these compounds, or the period required to adapt to degrade the compounds may be long. When the presence of chlorinated contaminants is posing environmental and health risks and little or no biodegradation of these compounds is detected, inoculation with contaminant-degrading microorganisms may be warranted.

11.2. MATURITY OF THE TECHNOLOGY

Inoculation or bioaugmentation has been widely used to stimulate bioremediation of subsurface material contaminated with petroleum hydrocarbons. A variety of cultures and formulations are commercially available. The practice of inoculation is based on the assumption that contamination has persisted in the subsurface because competent microorganisms were not available and that biodegradation is limited by active biomass. The practice further assumes that biodegradation of the contaminant is not limited by the supply of the substances required for metabolism of the contaminant, such as oxygen or mineral nutrients. Because adequate field evaluations have not been done, there is no way to determine whether perceived benefits were provided by the introduced organisms rather than indigenous organisms, or whether the effective agent was the introduced organism, or a mineral nutrient, surfactant, or biosurfactant provided along with the culture.

The inocula are relatively inexpensive compared to other remedial activities, and they are often used to insure the presence of a competent microbial community. If the inoculant is the only remedy, the contaminated site should be characterized to demonstrate that there are adequate supplies of electron acceptors and mineral nutrients to permit complete destruction of the contaminant.

The use of microorganisms with specialized capabilities to enhance bioremediation in the subsurface is not an established technology. However, research has been conducted to determine the potential for microbial transport through subsurface materials for public health and microbial enhanced oil recovery.

11.3. PRIMARY REPOSITORIES OF EXPERTISE

Institutions where microbial transport research is being conducted include Cornell University (M. Alexander), Mississippi State University (C. Hagedorn), Rice University (C. H. Ward), U. S. Geological Survey (R. W. Harvey), University of Arizona (C. Gerba), University of Calgary (J. W. Costerton), University of Idaho (R. L. Crawford), University of Oklahoma (G. E. Jenneman; M. J. McInerney), and University of Virginia (A. L. Mills). Current addresses and telephone numbers are listed below.

Martin Alexander
Department of Agronomy
Bradfield Hall
Cornell University
Ithaca, NY 14853
Phone #: (607)225-1717
FAX #: (607)255-2106

J. W. Costerton
Montana State University
Engineering Department
Center for Biofilm Engineering
409 Cobleigh Hall
Bozeman, MT 59717
Phone #: (406)994-4770
FAX #: (406)994-6098

Ronald L. Crawford
Food Research Center, Room 202
University of Idaho
Moscow, Idaho 83843
Phone #: (208)885-6580
FAX #: (208)885-5741

Charles Gerba
Department of Soil and Water Science
University of Arizona
Tuscon, AZ 85721
Phone #: (602)621-6906
FAX #: (602)621-1647

Charles Hagedorn
Department of Plant Pathology,
   Physiology, and Weed Science
Price Hall
Virginia Polytechnic Institute
Blacksburg, VA 24061
Phone #: (703)231-6361
FAX #: (703)231-7477

Ronald W. Harvey
U.S. Geological Survey
Water Resources Division
Box 25046, MS 458, Boulder Office
Denver, CO 80225
Phone #: (303)541-3034
FAX #: (303)447-2505

G. E. Jenneman
Phillips Petroleum Company
Bartlesville, OK 74005
Phone #: (918)661-8797
FAX #: (918)662-2047

Michael J. McInerney
Dept. of Botany and Microbiology
University of Oklahoma
Norman, OK 73019
Phone #: (405)325-6050
FAX #: (405)325-7619

Aaron Mills
Department of Environmental Sciences
Clark Hall
University of Virginia
Charlottesville, VA 22903
Phone #: (804)924-7761
FAX #: (804)982-2137

C. H. Ward
Department of Environmental
   Science and Engineering
Rice University
Houston, IX 77251
Phone #: (713)527-4086
FAX #: (713)285-5203

11.4. OTHER FACTORS CONCERNING APPLICATION

Although specialized microorganisms that have been cultured using selective enrichment techniques can be used in environmental applications, those developed using genetic engineering techniques cannot be released into the environment for commercial purposes without prior government approval (Pimentel et al., 1989). Genetically engineered microorganisms for use in such operations as MEOR, bioremediation of Superfund sites, extraction and concentration of metals, and production of specialty chemicals may be regulated under the Environmental Protection Agency's Toxic Substances Control Action Section 5 (Clark, 1992).

11.5. STATE OF THE ART OF TRANSPORT OF MICROORGANISMS WITH SPECIALIZED METABOLIC CAPABILITIES AND RESEARCH OPPORTUNITIES

Since the study published by Raymond et al. (1977) that indicated that microorganisms can be transported and enhance degradation of hydrocarbons in a column packed with sand, no one has demonstrated that inoculation of the subsurface can enhance bioremediation in the laboratory or field. There is a tendency to work with organisms that are easy to culture and whose genetics are well understood. Little consideration is given to developing organisms with good transport properties and survival traits. Provided that microorganisms can be successfully transported through a specified aquifer and establish themselves, several different possibilities for application exist (Table 11.1). The best opportunities involve development of inocula that can degrade mixed wastes, that have increased tolerance to toxicants, and that produce bioemulsifiers and biosurfactants to increase their access to oily phase contaminants.

TABLE 11.1. POSSIBLE APPLICATIONS OF INTRODUCED MICROORGANISMS

Specialized Capability	*Purpose*
Produce biosurfactant/bioemulsifier	Mobilize sorbed/entrained contaminants
Degrade multiple compounds	Treatment of mixture of compounds
Degrade recalcitrant compounds	Inoculation in absence of acclimation by indigenousorganisms
Tolerate and degrade toxic compound	Inoculation in absence of acclimation by indigenous organisms
Tolerate high concentration of toxicant	Inoculation in absence of acclimation by indigenous organisms

REFERENCES

Atlas, R.M. 1977. Stimulated petroleum biodegradation. *CRC Crit. Rev. Microbiol.* 5:371-386.

Beckmann, J.W. 1926. Action of bacteria on mineral oil. *Ind. Eng. Chem. News Ed.* 4:3.

Bossert, I., and R. Bartha. 1986. Structure-biodegradability relationships of polycyclic aromatic hydrocarbons in soil. *Bull. Environ. Contam. Toxicol.* 37:4490-4495.

Brunner, W., F.H. Sutherland, and D.D. Focht. 1985. Enhanced biodegradation of polychlorinated biphenyls in soil by analog enrichment and bacterial inoculation. *J. Environ. Qual.* 14(3):324-328.

Bubela, B. 1978. Role of geomicrobiology in enhanced recovery of oil: status quo. *APEA J.* B18:161-166.

Caldwell, E.L. 1937. Pollution flow from pit latrines when an impervious stratum closely underlies the flow. *J. Infect. Dis.* 61:269-288.

Caldwell, E.L. 1938. Studies of subsoil pollution in relation to possible contamination of the ground water from human excreta deposited in experimental latrines. *J. Infect. Dis.* 62:273-292.

Clark, E. 1992. EPA's coverage of bioremediation activities under TSCA. *Biotreatment News.* 2:8.

Crawford, R.L., and W.W. Mohn. 1985. Microbial removal of pentachlorophenol from soil using a *Flavobacterium. Enzyme Microb. Technol.* 7:617-620.

Edgehill, R.U., and R.K. Finn. 1983. Microbial treatment of soil to remove pentachlorophenol. *Appl. Environ. Microbiol.* 45(3):1122-1125.

Flathman, P.E., and J.A. Caplan. 1985. Biological cleanup of chemical spills. I *Proceedings of HAZMACON 85.* Association of Bay Area Governments. Oaklar California. pp. 323-345.

Flathman, P.E., and J.A. Caplan. 1986. Cleanup of contaminated soils and grou water using biological techniques. In: *Proceedings National Conference Hazardous Wastes and Hazardous Materials.* Hazardous Materials Cont Research Institute. Silver Spring, Maryland. pp. 110-119.

Flathman, P.E., and G.D. Githens. 1985. In situ biological treatment of isopropan acetone, and tetrahydrofuran in the soil/ground water environment.   ] *Groundwater Treatment Technology.* Ed., E.K. Nyer. Van nostrand Reinh(Company. New York. pp. 173-185.

Flathman, P.E., M.J. McCloskey, J.J. Vondrick, and D.W. Pimlett. 1985. In s physical/biological treatment of methylene chloride (dichlorometha) contaminated ground water. In: *Proceedings Fifth National Symposium Aquifer Restoration and Ground Water Monitoring.* National Water W Association. Worthington, Ohio. pp. 571-597.

Fontes, D. E., A.L. Mills, G.M. Hornberger, and J.S. Sherman. 1991. Physical and chemical factors influencing transport of microorganisms through porous media. *Appl. Environ. Microbiol.* 57(9):2473-2481.

Gannon, J.T., V.B. Manilal, and M. Alexander. 1991a. Relationship between cell surface properties and transport of bacteria through soil. *Appl. Environ. Microbiol.* 57(1):190-193.

Gannon, J., Y. Tan, P. Baveye, and M. Alexander. 1991b. Effect of sodium chloride on transport of bacteria in a saturated aquifer material. *Appl. Environ. Microbiol.* 57(9):2497-2501.

Gerba, C.P., and G. Bitton. 1984. Microbial pollutants: their survival and transport pattern to groundwater. In: *Groundwater Pollution Microbiology.* Eds., G. Bitton and C.P. Gerba. John Wiley and Sons, Inc. New York. pp. 65-88.

Goldstein, R.M., L.M. Mallory, and M. Alexander. 1985. Reasons for possible failure of inoculation to enhance biodegradation. *Appl. Environ. Microbiol.* 50(4):977-983.

Hagedorn, C., D.T. Hansen, and G.H. Simonson. 1978. Survival and movement of fecal indicator bacteria in soil under conditions of saturated flow. *J. Environ. Qual.* 7(1):55-59.

Hart, R.T., T. Fekete, and D.L. Flock. 1960. The plugging effect of bacteria in sandstone systems. *Can. Min. Metall. Bull.* 53:495-501.

Harvey, R.W., L.H. George, R.L. Smith, and D.R. LeBlanc. 1989. Transport of microspheres and indigenous bacteria through a sandy aquifer: results of natural- and forced-gradient tracer experiments. *Environ. Sci. Technol.* 23(1):51-56.

Heitkamp, M.A., and C.E. Cerniglia. 1988. Mineralization of polycyclic aromatic hydrocarbons by a bacterium isolated from a sediment below an oil field. *Appl. Environ. Microbiol.* 54(6):1612-1614.

Jain, R.K., G.S. Sayler, J.T. Wilson, L. Houston, and D. Pacia. 1987. Maintenance and stability of introduced genotypes in groundwater aquifer material. *Appl. Environ. Microbiol.* 53(5): 996-1002.

Jang, L-K., P.W. Chang, J.E. Findley, and T.F. Yen. 1983. Selection of bacteria with favorable transport properties through porous rock for the application of microbial-enhanced oil recovery. *Appl. Environ. Microbiol.* 46(5): 1066-1072.

Jenneman, G.E., R.M. Knapp, M.J. McInerney. 1985. Microbial penetration through nutrient-saturated Berea sandstone. *Appl. Environ. Microbiol.* 50(2):383-391.

Jobson, A., M. McLaughlin, F.D. Cook, and D.W.S. Westlake. 1974. Effect of amendments on the microbial utilization of oil applied to soil. *Appl. Microbiol.* 27:166-171.

Kalish, P.J., J.A. Stewart, W.F. Rogers, and E.O. Bennett. 1964. The effect of bacteria on sandstone permeability. *J. Petrol. Technol.* 16:805-814.

Kelley, I., and C.E. Cerniglia. 1991. The metabolism of fluoranthrene by a species of *Mycobacterium*. *J. Indus. Microbiol.* 7:19-26.

Lee, M. D., J.M. Thomas, R.C. Borden, P.B. Bedient, J.T. Wilson, and C.H. Ward. 1988. Biorestoration of aquifer contaminated with organic compounds. *CRC Crit. Rev. Environ. Control.* 18: 29-89.

Lehtomaki, M., and S. Niemela. 1975. Improving microbial degradation of oil in soil. *Ambio.* 4:126-129.

MacLeod, F.A., H.M. Lappin-Scott, and J.W. Costerton. 1988. Plugging of a model rock system by using starved bacteria. *Appl. Environ. Microbiol.* 54(6):1365-1372.

MacRae, I.C., and M. Alexander. 1965. Microbial degradation of selected herbicides in soil. *J. Agric. Food Chem.* 13:72-76.

Madsen, E.L., and M. Alexander. 1982. Transport of *Rhizobium* and *Pseudomonas* through soil. *Soil Sci. Soc. Am. J.* 46:557-560.

Madsen, E.L., J.L. Sinclair, and W.C. Ghiorse. 1991. In situ biodegradation: microbiological patterns in a contaminated aquifer. *Science.* 252(5007):830-833.

Marlow, H.J., K.L. Duston, M.R. Wiesner, M.B. Tomson, J.T. Wilson, and C.H. Ward. 1991. Microbial transport through porous media: the effects of hydraulic conductivity and injection velocity. *J. Hazard. Mat.* 28:65-74.

Martinson, M.M., J.G. Steiert, D.L. Saber, and R.L. Crawford. 1984. Microbiological decontamination of pentachlorophenol in natural waters. In: *Biodeterioration Society: Proceedings of the Sixth International Symposium.* Ed., E.E. O'Rear. Commonwealth Agricultural Bureau.

McCoy, E.L., and C. Hagedorn. 1980. Transport of resistance-labeled *Escherichia coli* strains through a transition between two soils in a topographic sequence. *J. Environ. Qual.* 9(4):686-691.

Mueller, J.G., P.J. Chapman, B.O. Blattmann, and P.H. Pritchard. 1990. Isolation and characterization of a fluoranthrene-utilizing strain of Pseudomonas paucimobilis. *Appl. Environ. Microbiol.* 56(4):1079-1086.

Ohneck, R.J., and G.L. Gardner. 1982. Restoration of an aquifer contaminated by an accidental spill of organic chemicals. *Ground Water Monitoring Review.* 2(4):50-53.

Pimentel, D., M.S. Hunter, J.A. LaGro, R.A. Efroymson, J.C. Landers, F.T. Mervis, C.A. McCarthy, and A.E. Boyd. 1989. Benefits and risks of genetic engineering in agriculture. *Bioscience.* 39:606-614.

Quince, J.R., and G.L. Gardner. 1982a. Recovery and treatment of contaminated ground water, Part I. *Ground Water Monitoring Review.* 2(3):18-22.

Quince, J.R., and G.L. Gardner. 1982b. Recovery and treatment of contaminated ground water, Part II. *Ground Water Monitoring Review.* 2(4):18-25.

Quince, J.R., R.J. Ohneck, and J.J. Vondrick. 1985. Response to an environmental incident affecting ground water. In: *Proceedings Fifth National Symposium and*

Exposition on Aquifer Restoration and Ground Water Monitoring. National Water Well Association. Worthington, Ohio. pp. 598-608.

Rahe, T.M., C. Hagedorn, and G.F. Kling. 1978a. Transport of antibiotic-resistant *Escherichia coli* through western Oregon hillslope soils under conditions of saturated flow. *J. Environ. Qual*. 7(4):487-494.

Rahe, T.M., C. Hagedorn, and E.L. McCoy. 1978b. A comparison of fluorescein dye and antibiotic-resistant *Escherichia coli* as indicators of pollution in groundwater. *Water, Air, and Soil Pollution.* 11:93-103.

Ramadan, M.A., O.M. El-Tayeb, and M. Alexander. 1990. Inoculum size as a factor limiting success of inoculation for biodegradation. *Appl. Environ. Microbiol.* 56(5):1392.

Raymond, R.L., J.O. Hudson, and V.W. Jamison. 1977. *Bacterial Growth in and Penetration of Consolidated and Unconsolidated Sands Containing Gasoline.* API Publication No. 4426. American Petroleum Institute. Washington, DC.

Reynolds, P.J., P. Sharma, G.E. Jenneman, and M.J. McInerney. 1989. Mechanisms of microbial movement in subsurface materials. *Appl. Environ. Microbiol*. 55(9): 2280-2286.

Scholl, M.A., A.L. Mills, J.S. Herman, and G.M. Hornberger. 1990. The influence of mineralogy and solution chemistry on the attachment of bacteria to representative aquifer materials. *J. Contam. Hydrol.* 6:321-336.

Schwendinger, R. B. 1968. Reclamation of soil contaminated with oil. *J. Instit. Petrol*. 54:182-197.

Sinclair, J. L. 1991. Eucaryotic microorganisms in subsurface environments. In: *Proceedings of the First International Symposium on Microbiology of the Deep Subsurface.* Eds., C.B. Fliermans and T.C. Hazen. Jan. 15-19, 1990. Orlando Florida. WSRC Information Services. Aiken, South Carolina. pp. 3-39 - 3-48.

Smith, M. S., G.W Thomas, and R.E. White. 1983. *Movement of Bacteria Through Macropores to Ground Water.* Research Report No. 139. United States Department of the Interior. University of Kentucky, Water Resources Research Institute. Lexington, Kentucky. 33 pp.

Stormo, K.E., and R.L. Crawford. 1992. Preparation of encapsulated microbial cells for environmental applications. *Appl. Environ. Microbiol*. 58(2):727-730.

Thomas, J.M., and C.H. Ward. 1989. In situ biorestoration of organic contaminants in the subsurface. *Environ. Sci. Tech*. 23(7):760-766.

Westlake, D.W.S, A.M. Jobson, and F.D. Cook. 1978. In situ degradation of oil in a soil of the boreal region of the Northwest Territories. *Can. J. Microbiol*. 24:254-260.

Winegardner, D.L., and J.R. Quince. 1984. Ground water restoration projects: five case histories. In: *Proceedings Fourth National Symposium and Exposition on Aquifer Restoration and Ground Water Monitoring.* National Water Well Association. Worthington, Ohio. pp. 386-393.

Zaidi, B.R., Y. Murakami, and M. Alexander.   1989.   Predation and inhibitors in lake water affect the success of inoculation to enhance biodegradation of organic chemicals.   *Environ. Sci. Technol.*   23(7):859-863.

Zaidi, B.R., G. Stucki, and M. Alexander.   1988.   Low chemical concentration and pH as factors limiting the success of inoculation to enhance biodegradation.   *Environ. Toxicol. Chem.*   7:143-151.

Zajic, T.E., and A.J. Daugulis.   1975.   Selective enrichment processes in resolving hydrocarbon pollution problems.   In: *Proceedings, Impact of the Use of Microorganisms on the Aquatic Environment.*   EPA 660-3-75-001.   U.S. Environmental Protection Agency. Corvallis, Oregon.   pp. 169-182.

INDEX